Nutraceutical Applications in Food Processing and Preservation

Nutraceutical Applications in Food Processing and Preservation

Edited by

M. Anuradha

Padmashree Institute of Management and Sciences, Bengaluru, Karnataka, India

S. Balasubramanya

Association for Biotechnology Led Enterprises (ABLE), Bengaluru, Karnataka, India

ELSEVIER

ACADEMIC PRESS

An imprint of Elsevier

Academic Press is an imprint of Elsevier
125 London Wall, London EC2Y 5AS, United Kingdom
50 Hampshire Street, 5th Floor, Cambridge, MA 02139, United States

ISBN: 978-0-443-45784-5

For Information on all Academic Press publications
visit our website at https://www.elsevier.com/books-and-journals

Publisher: Mara Conner
Acquisitions Editor: Vaishali Singh
Editorial Project Manager: Sadhna Rajan
Production Project Manager: Aishwarya Ramesh
Cover Designer: Mark Rogers

Typeset by MPS Limited, Chennai, India

Contents

List of contributors

M. Anuradha Padmashree Institute of Management and Sciences, Bengaluru, Karnataka, India

S. Balasubramanya Association for Biotechnology Led Enterprises (ABLE), Bengaluru, Karnataka, India

Chandan Hebbale Brown University, Providence, RI, United States; Brookside Botanical Gardens, Takoma Park, MD, United States

B.K. Indu Department of Biotechnology, Padmashree Institute of Management and Sciences, Bengaluru, Karnataka, India

Nagamani J.E. Department of Food Science and Nutrition, Padmashree Institute of Management and Sciences, Bengaluru, Karnataka, India

Karthik Maniam SRK Nutritionals, LLC, Ronkonkoma, NY, United States

Yamuna Mannava SRK Nutritionals, Ronkonkoma, NY, United States

Sudipta Kumar Mohanty Department of Biotechnology, Padmashree Institute of Management and Sciences, Bengaluru, Karnataka, India

Shilpa P. Department of Nutrition and Dietetics, Padmashree Institute of Management and Sciences, Bengaluru, Karnataka, India

M.R. Raghunath Department of Food Science and Technology, Padmashree Institute of Management and Sciences, Bengaluru, Karnataka, India

Velumurugan Shanmugam Business Development & Quality Affairs, Biogenero Labs Pvt. Ltd., Bengaluru, Karnataka, India; Phytosol India Pvt. Ltd., Bengaluru, Karnataka, India

Umalatha Department of Food Science and Technology, Padmashree Institute of Management and Sciences, Bengaluru, Karnataka, India

Preface

Food processing has always been fundamental to human survival and progress, ensuring food security, reducing post-harvest losses, extending shelf life, and enhancing the value of agricultural produce. In today's context of rapid population growth, urbanization, and changing lifestyles, the food processing sector has become indispensable in bridging the gap between farm and consumer, while also contributing significantly to farmer income and economic development. However, alongside these advancements, there has been a growing concern regarding the extensive use of synthetic additives, preservatives, artificial colors, and flavoring agents in processed foods. Modern consumers are increasingly aware and selective, seeking foods that are not only safe and convenient but also natural, nutritious, and health-promoting. This evolving demand highlights a critical shift from mere food security to a more comprehensive goal of nutrition security and healthy food systems.

In this transition, nutraceuticals have emerged as a powerful and promising solution. Derived from natural sources, nutraceuticals offer dual functionality, providing health benefits while also playing technological roles in food processing and preservation. They represent a paradigm shift in the way food is designed, processed, and consumed, enabling the replacement of synthetic additives with natural, functional, and safer alternatives. The integration of nutraceuticals into food systems aligns with the global movement toward clean-label products, sustainable processing, and preventive healthcare through diet.

This book, *Nutraceutical Applications in Food Processing and Preservation*, is developed with the objective of addressing this critical need by bringing together scientific knowledge and practical applications in a cohesive manner. It presents a comprehensive exploration of nutraceuticals, beginning with their fundamentals and progressing toward their application as natural alternatives to conventional food additives such as preservatives, sweeteners, colorants, emulsifiers, and flavoring agents. The chapters further extend into industry-specific applications, safety and quality considerations, consumer trends, and future perspectives, providing a holistic understanding of the field.

The scientific significance of this work lies in its interdisciplinary approach, integrating concepts from food science, nutrition, biochemistry, and biotechnology to elucidate the mechanisms, functionality, and potential of nutraceutical ingredients. Technically, it addresses formulation strategies, challenges in clean-label product development, quality assurance, and regulatory aspects, offering practical insights for both academia and industry. Importantly, this book also identifies research gaps and future directions, emphasizing the need for innovation in natural ingredient development, bioavailability enhancement, and scalable processing technologies.

The scope of this book is not only to inform but also to inspire. It aims to guide researchers, academicians, industry professionals, and policymakers toward developing healthier, safer, and more sustainable food systems. By focusing on the replacement of unhealthy additives with nutraceutical alternatives and promoting health-oriented processing approaches, this work aspires to contribute to a future where food processing supports both human health and environmental sustainability.

Introduction to nutraceuticals in food processing

1

M. Anuradha[1] *and S. Balasubramanya*[2]
[1]Padmashree Institute of Management and Sciences, Bengaluru, Karnataka, India,
[2]Association for Biotechnology Led Enterprises (ABLE), Bengaluru, Karnataka, India

1.1 Food as medicine: from philosophy to practice

Food systems today face the dual challenge of ensuring nutritional adequacy while maintaining safety, sustainability, and accessibility across diverse populations. Within this context, the longstanding principle that food itself can contribute to health and disease prevention has gained renewed scientific and societal relevance.

"Let food be thy medicine" is a timeless philosophy that continues to resonate across cultures and generations. This simple yet profound statement captures the fundamental role of food not merely as a source of sustenance, but as a powerful determinant of health, vitality, and longevity. Long before the advent of modern medicine, traditional societies recognized that daily dietary choices could influence physical strength, resistance to illness, and overall well-being. Food was viewed as the first line of defense against disease and the foundation of a healthy life.

In contemporary times, this traditional wisdom has found strong validation through advances in nutrition science and food technology. It is now well established that diet plays a critical role in the prevention and management of a wide range of health conditions, including metabolic disorders, cardiovascular diseases, micronutrient deficiencies, and lifestyle-related ailments (Granato et al., 2020; Martirosyan and Singh, 2015). Food supplies essential nutrients, bioactive compounds, and functional components that collectively support physiological functions and maintain homeostasis. Thus, food remains central to health promotion and disease prevention, reinforcing its role as a form of everyday medicine.

However, the health potential of food does not depend solely on the nature of raw agricultural materials. Equally important are the ways in which foods are prepared, processed, and preserved before consumption. From harvesting and handling to cooking and storage, each step has the capacity to influence the nutritional quality, safety, and functional value of food (Fellows, 2017). Improper practices can lead to significant losses of sensitive nutrients and bioactive compounds, while scientifically designed processes can help retain, stabilize, or even enhance their availability.

This evolving understanding marks a transition from viewing processing merely as a means of extending shelf life to recognizing it as a critical intervention that shapes the medicinal value of food. Processing technologies can either erode the inherent health benefits of natural foods or serve as tools to protect and optimize them. When guided by sound scientific principles, they have the potential to preserve nutritional integrity, improve bioavailability, and support the development of foods that actively

Nutraceutical Applications in Food Processing and Preservation. DOI: https://doi.org/10.1016/B978-0-443-45784-5.00013-7

contribute to health and wellness (Kearney, 2010; Tiwari and Gowen, 2016). This perspective forms the foundation for exploring the role of processing and preservation within modern food systems, where the challenge lies in harmonizing safety, convenience, nutrition, and health promotion.

1.1.1 Food preparation and processing versus preservation of nutritional and functional value

Food preparation and processing play a decisive role in determining the nutritional and functional quality of foods consumed daily. While raw foods provide the primary matrix of nutrients, their actual contribution to human health depends on how effectively these nutrients are retained and made available during preparation, processing, and subsequent consumption. Simple household practices such as washing, cutting, soaking, and cooking, as well as industrial operations like milling, drying, thermal treatment, and packaging, can significantly influence nutrient retention and bioavailability (Fellows, 2017).

Many essential nutrients, particularly vitamins, antioxidants, and other bioactive compounds, are highly sensitive to heat, light, oxygen, and mechanical stress. Improper handling and unscientific processing practices can therefore result in substantial nutrient losses, diminishing the health value of foods even before they reach the consumer (Tiwari and Gowen, 2016). Excessive thermal treatments, prolonged exposure to air, and poor storage conditions may lead to degradation of vitamins, oxidation of lipids, and loss of functional compounds, thereby reducing both nutritional density and physiological efficacy.

At the same time, processing should not be viewed solely as a cause of nutrient loss. When applied as a controlled and scientifically guided intervention, it serves as a powerful tool to ensure food safety, improve shelf life, and maintain overall quality. Techniques such as blanching, fermentation, controlled drying, and appropriate thermal treatments can inactivate harmful microorganisms and enzymes while preserving essential nutrients (Rahman, 2007). Moreover, certain processes enhance nutrient bioavailability by breaking down complex food matrices, reducing antinutritional factors, and improving digestibility.

Importantly, modern food technologies open avenues for functional enhancement through thoughtful formulation. By selecting suitable processing approaches and combining them with natural or bioactive ingredients, it is possible to design foods that not only retain their inherent nutritional value but also offer added health benefits (Granato et al., 2020). This shift from conventional processing to value-oriented processing highlights the potential of food systems to deliver products that support wellness, bridge nutritional gaps, and align with the growing demand for health-focused diets.

1.1.2 Need for processing and preservation

Fresh agricultural produce is inherently perishable. Most fruits, vegetables, dairy products, and fish begin to deteriorate within hours of harvest if not handled appropriately. Microbial growth, moisture loss, enzymatic changes, and mechanical damage during handling and transport can rapidly diminish both the quantity and quality of foods intended for consumption (Fellows, 2017).

In many parts of the world, including India, this vulnerability becomes especially evident during seasonal harvest peaks. Crops such as tomatoes, mangoes, bananas, and leafy vegetables are often produced in large quantities during short time windows, leading to temporary oversupply. Without adequate preservation and storage infrastructure, a significant portion of this seasonal abundance may be lost before reaching consumers (Fig. 1.1).

The magnitude of food loss and waste is considerable. Globally, approximately one-third of food produced for human consumption, equivalent to about 30%–40%, is lost or wasted across the food supply chain (Gustavsson et al., 2011; FAO, 2011). Perishable commodities such as fruits and vegetables are particularly vulnerable due to their high moisture content and sensitivity to environmental conditions.

In the Indian context, the challenge is similarly significant. National assessments indicate that cereals such as paddy and wheat may experience post-harvest losses of approximately 4%–6%, while fruits and vegetables may experience losses ranging from 6% to 15% or higher, largely due to limitations in storage, transportation, and cold-chain infrastructure (MoFPI Ministry of Food Processing Industries, 2015; FAO, Government of India, 2014). When aggregated across the entire food system, these losses translate into tens of millions of tons of food annually.

Food losses are multilayered in nature. At the field level, delays in harvesting or improper threshing can result in grain losses. During transportation and storage, inadequate packaging, temperature control, and infrastructure accelerate deterioration. Inefficiencies within wholesale and retail markets further compound these losses. Beyond the nutritional and economic consequences, the resources embedded in lost food, land, water, energy, and labor are also wasted. Food that is produced but never consumed represents a substantial environmental burden and contributes to unnecessary pressure on natural resources (HLPE, 2014; FAO, 2011). These realities highlight the essential role of preservation and processing interventions in stabilizing food supplies, reducing losses, and ensuring that agricultural production translates into real nutritional benefit.

1.1.3 Food processing and preservation as pillars of food security

Food security extends far beyond the mere production of sufficient quantities of food. It encompasses four interconnected dimensions: availability, access, stability, and utilization (FAO, 2008). In addition, nutrition security represents an equally important dimension, emphasizing the quality of diets and the adequate intake of essential nutrients required for maintaining health and leading a healthy life. Processing contributes directly to availability by transforming highly perishable agricultural products into stable forms that can be stored and transported over longer periods. Cereals milled into flour, fruits processed into pulps or dried products, milk converted into dairy products, and meat preserved through chilling or curing all

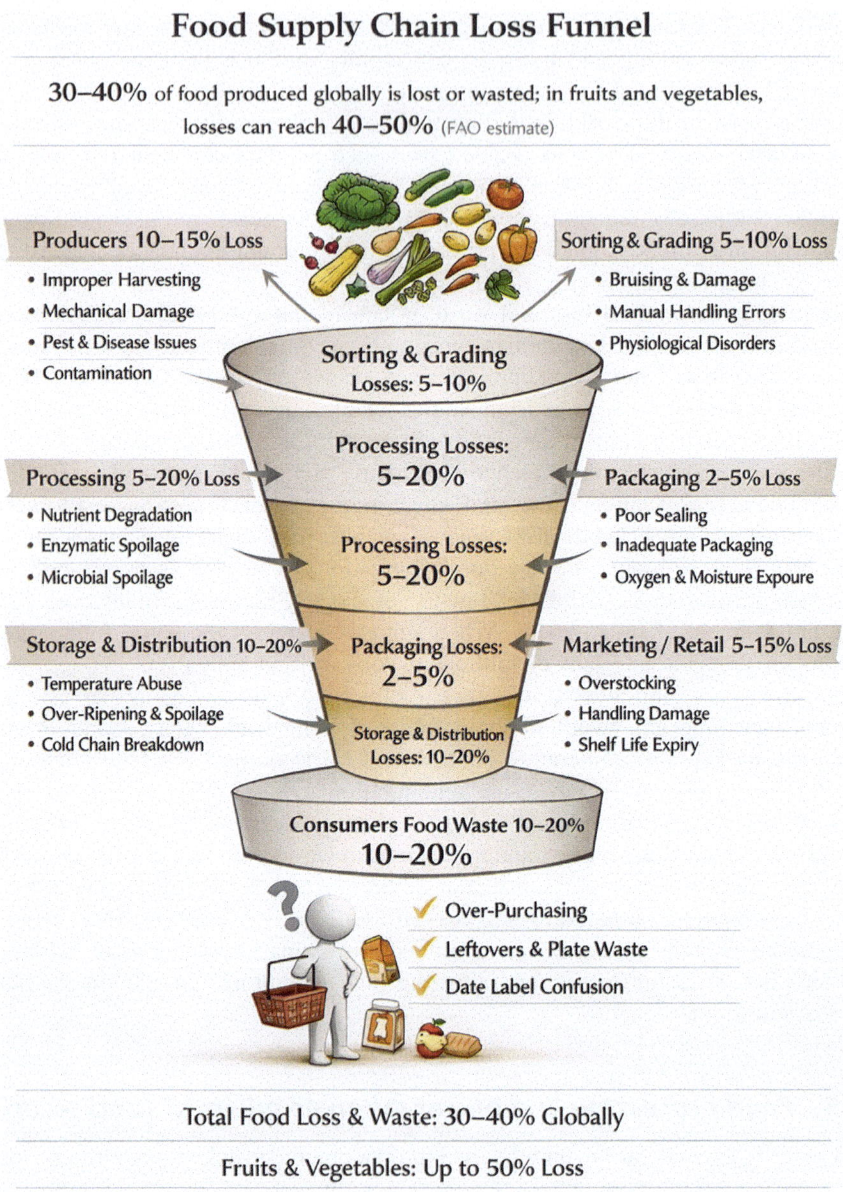

Figure 1.1 Food supply chain loss. The figure shows the food supply chain loss funnel, indicating that about 30%–40% of global food production is lost or wasted, with fruits and vegetables losing up to 50%. Losses occur across stages such as harvesting, processing, storage, retail, and consumer handling due to damage, spoilage, poor storage, and waste.

illustrate how technological interventions extend the usable life of food commodities (Fellows, 2017).

Accessibility is enhanced when foods can be distributed across geographical regions and supplied to diverse population groups. Preservation technologies allow food to be transported over long distances and stored within modern retail systems, reducing dependence on local production cycles. Equally important is the role of preservation strategies in ensuring the stability of supply. Agricultural production is inherently vulnerable to climatic variability, seasonal fluctuations, and unforeseen disruptions. By converting seasonal surpluses into preserved forms, processing helps buffer fluctuations in production and ensures availability during lean periods.

The fourth dimension, utilization, relates to the body's ability to derive nutritional value from food. Safe processing practices eliminate pathogenic microorganisms, reduce contamination risks, and improve digestibility (Rahman, 2007). Through these mechanisms, processing ensures that food not only reaches consumers but also contributes effectively to nutritional well-being. Thus, processing and preservation function as critical links between agricultural production and nutritional security.

1.1.4 Natural and nutraceutical additives: intent, rationale, and promise

While conventional preservation strategies have traditionally relied on physical and chemical methods, increasing attention is now being directed toward biologically derived ingredients that simultaneously support preservation and health promotion. The growing reliance on processed foods has historically been accompanied by the use of synthetic additives to enhance shelf life, improve appearance, and ensure product stability. However, increasing consumer awareness has led to concerns about artificial preservatives, colorants, and flavor enhancers (Asioli et al., 2017).

As a result, there has been a clear shift toward clean-label food products characterized by simplicity, transparency, and recognizable ingredients. Natural additives and nutraceutical components derived from biological sources, including plant extracts, bioactive compounds, fermentation products, and natural antioxidants, are increasingly being explored as alternatives to conventional additives (Granato et al., 2020). These ingredients provide multifunctional benefits. In addition to contributing to preservation through antimicrobial and antioxidant properties, they may also enhance nutritional value and support physiological health (Fig. 1.2). This evolving approach reflects the concept of "preservation with purpose," in which food safety, shelf-life stability, and nutritional enhancement are addressed simultaneously rather than in isolation.

1.1.5 Benefits to the food industry

The incorporation of natural and nutraceutical additives provides several strategic advantages to the food industry. In a competitive marketplace where consumers increasingly value transparency and health-oriented products, such ingredients offer opportunities for meaningful product differentiation. Clean-label formulations that

Figure 1.2 Processing and preservation with purpose framework. Framework illustrating the multidimensional role of food processing in modern food systems. Scientifically guided processing supports nutrient retention, ensures food safety and quality, reduces food waste, and enhances functional and health benefits, thereby contributing to sustainable food systems and improved public health.

replace or reduce synthetic additives are often perceived as safer and more trustworthy by consumers (Asioli et al., 2017). Natural antioxidants and antimicrobial compounds can extend shelf life while maintaining desirable sensory attributes such as taste, aroma, and texture. From a market perspective, foods enriched with natural bioactive compounds align closely with global trends toward functional foods and preventive nutrition (Granato et al., 2020; Martirosyan and Singh, 2015). Such products enable manufacturers to respond to evolving consumer expectations while maintaining regulatory compliance and product quality.

1.1.6 Benefits to individuals and public health

The integration of nutraceutical components into commonly consumed foods has important implications for individual health and public nutrition. Bioactive compounds such as polyphenols, peptides, phytochemicals, and natural antioxidants may contribute to improved metabolic health, reduced oxidative stress, and enhanced physiological resilience (Granato et al., 2020). Foods formulated with such components can help address micronutrient deficiencies and support the management of lifestyle-related disorders, including cardiovascular diseases and metabolic syndromes. Importantly, delivering bioactive compounds through everyday foods encourages sustainable dietary habits. Instead of relying solely on supplements or specialized dietary regimens, consumers can obtain health-promoting compounds through regular dietary patterns.

1.1.7 Advantages to farmers and the agrifood ecosystem

The adoption of nutraceutical-oriented processing strategies also presents significant opportunities for farmers and the broader agrifood ecosystem. Agricultural by-products such as peels, seeds, pomace, and husks are increasingly recognized as valuable sources of bioactive compounds (Shahidi and Ambigaipalan, 2015). The utilization of such materials for functional ingredient production promotes resource efficiency and supports the principles of a circular bioeconomy. In addition, value-added processing can increase farm income, reduce post-harvest losses, and encourage rural entrepreneurship through decentralized food processing enterprises. These developments strengthen linkages between agriculture, food processing, and nutrition, contributing to a more resilient and sustainable food system.

1.1.8 Scope of the chapter

This chapter provides an integrated overview of the principles and evolving role of food processing and preservation in contemporary food systems. It examines how these interventions influence food safety, shelf life, quality, and nutritional value. Particular emphasis is placed on the emerging role of natural and nutraceutical additives in modern food processing. The chapter explores their potential to extend shelf life, protect sensitive nutrients, and enhance the health value of foods while meeting consumer expectations for natural and clean-label products. The chapters that follow in this book build upon this foundation by exploring the science, technology, and applications of natural bioactive compounds and nutraceutical ingredients in food systems. Topics include sources of functional ingredients, preservation strategies, formulation approaches, and emerging technologies that support the development of safe, stable, and health-oriented food products.

1.1.9 Contemporary relevance and significance

The relevance of food processing and preservation has intensified in recent years due to profound changes in dietary patterns, health awareness, and global food system dynamics. Consumers increasingly seek foods that contribute positively to well-being rather than merely satisfying caloric needs. In response, the food industry is undergoing a transition toward functional, clean-label, and nutritionally enriched products (Granato et al., 2020).

At the same time, global concerns related to food loss, sustainability, and nutritional insecurity highlight the importance of efficient preservation technologies (FAO, 2011; HLPE, 2014). Within this broader context, food processing should be viewed not merely as a technological necessity but as a strategic intervention that links agricultural production, nutritional quality, and public health. When guided by scientific innovation and sustainability principles, processing can play a transformative role in building resilient food systems capable of nourishing populations in an equitable and sustainable manner.

AI disclosure

During the preparation of this work the author(s) used open AI tool. Artificial intelligence tools were used only for minor language editing, including grammar correction, sentence refinement, and improvement of readability. After using this tool/service, the author(s) reviewed and edited the content as needed and take(s) full responsibility for the content of the publication.

References

Asioli, D., Aschemann-Witzel, J., Caputo, V., Vecchio, R., Annunziata, A., Næs, T., et al., 2017. Making sense of the "clean label" trends: a review of consumer food choice behavior and discussion of industry implications. Food Research International 99 (Pt 1), 58–71. Available at https://doi.org/10.1016/j.foodres.2017.07.022.

FAO, 2008. An Introduction to the Basic Concepts of Food Security. Food and Agriculture Organization of the United Nations, Rome.

FAO, 2011. Global Food Losses and Food Waste: Extent, Causes and Prevention. Food and Agriculture Organization of the United Nations, Rome.

FAO, Government of India, 2014. Food Losses in India: Extent, Causes and Policy Options. Food and Agriculture Organization of the United Nations, Rome.

Fellows, P.J., 2017. Food Processing Technology: Principles and Practice, fourth ed. Woodhead Publishing, Cambridge.

Granato, D., Barba, F.J., Bursać Kovačević, D., Lorenzo, J.M., Cruz, A.G., Putnik, P., 2020. Functional foods: product development, technological trends, efficacy testing, and safety. Annual Review of Food Science and Technology 11, 93–118. Available at https://doi.org/10.1146/annurev-food-032519-051708, http://www.annualreviews.org/journal/food.

Gustavsson, J., Cederberg, C., Sonesson, U., van Otterdijk, R., Meybeck, A., 2011. Global Food Losses and Food Waste: Extent, Causes and Prevention. Food and Agriculture Organization of the United Nations, Rome.

HLPE, 2014. Food Losses and Waste in the Context of Sustainable Food Systems. A Report by the High Level Panel of Experts on Food Security and Nutrition of the Committee on World Food Security, Rome.

Kearney, J., 2010. Food consumption trends and drivers. Philosophical Transactions of the Royal Society B: Biological Sciences 365 (1554), 2793–2807. Available at https://doi.org/10.1098/rstb.2010.0149.

Martirosyan, D.M., Singh, J., 2015. A new definition of functional food by FFC: What makes a new definition unique? Functional Foods in Health and Disease 5 (6), 209–223. Available at https://doi.org/10.31989/ffhd.v5i6.183, https://ffhdj.com/index.php/ffhd/article/view/183.

MoFPI (Ministry of Food Processing Industries), 2015. Assessment of Quantitative Harvest and Post-Harvest Losses of Major Crops and Commodities in India. Government of India, New Delhi.

Rahman, M.S., 2007. Handbook of Food Preservation, second ed. CRC Press, Boca Raton.

Shahidi, F., Ambigaipalan, P., 2015. Phenolics and polyphenolics in foods, beverages and spices: antioxidant activity and health effects—a review. Journal of Functional Foods 18, 820–897. Available at https://doi.org/10.1016/j.jff.2015.06.018.

Tiwari, B.K., Gowen, A., 2016. Nonthermal Processing Technologies for Food. Elsevier Academic Press, London.

Fundamentals of nutraceuticals

2

M. Anuradha[1] *and S. Balasubramanya*[2]
[1]Padmashree Institute of Management and Sciences, Bengaluru, Karnataka, India,
[2]Association for Biotechnology Led Enterprises (ABLE), Bengaluru, Karnataka, India

2.1 Introduction

Food has traditionally been viewed as a source of energy and essential nutrients required for growth, development, and maintenance of normal physiological functions. However, advances in nutrition science have clearly demonstrated that foods can provide benefits that extend beyond basic nourishment. Certain foods and food components play an important role in promoting health, reducing the risk of chronic diseases, and supporting overall physiological well-being. This evolving understanding has led to the emergence of the concept of nutraceuticals.

Nutraceuticals represent a rapidly growing field that integrates food science, nutrition, biochemistry, and health sciences. They are increasingly recognized for their role in preventive nutrition, particularly in the management of lifestyle-related disorders such as cardiovascular diseases, diabetes, obesity, and certain cancers. As a result, nutraceuticals have gained significant attention from researchers, healthcare professionals, and the food industry (Ross et al., 2020).

2.1.1 Definition of nutraceuticals

The term nutraceutical is derived from the words nutrition and pharmaceutical, reflecting the dual role of these products as food-derived substances with health-promoting properties. Nutraceuticals are broadly defined as foods or food components that provide medical or health benefits, including the prevention and management of disease. Watson (2015) describes nutraceuticals as isolated nutrients, dietary supplements, herbal products, functional foods, and bioactive compounds that exert physiological effects beyond basic nutritional functions. Unlike pharmaceutical drugs, nutraceuticals are generally consumed as part of the daily diet or in supplement form and are intended to support normal body functions rather than to cure diseases. Importantly, nutraceuticals occupy an intermediate position between conventional foods and medicines. Their health benefits are typically realized through regular and long-term consumption, contributing to health maintenance and reduction of disease risk rather than immediate therapeutic action (Bagchi and Preuss, 2013).

Nutraceutical Applications in Food Processing and Preservation. DOI: https://doi.org/10.1016/B978-0-443-45784-5.00003-4

2.1.2 Scope of nutraceuticals

The scope of nutraceuticals is extensive and continues to expand due to advances in food chemistry, biotechnology, and nutritional research. Nutraceuticals include a wide range of substances such as:

- Vitamins and minerals with enhanced physiological roles
- Bioactive peptides derived from dietary proteins
- Essential fatty acids such as omega-3 and omega-6 fatty acids
- Phytochemicals including polyphenols, flavonoids, and carotenoids
- Probiotics and prebiotics that modulate gut microbiota

These compounds influence various biological processes including antioxidant defense, immune modulation, lipid metabolism, glucose regulation, and gastrointestinal health (Ross et al., 2020). Nutraceuticals are increasingly incorporated into products designed for specific population groups such as infants, elderly individuals, athletes, and individuals with special nutritional requirements. Consequently, nutraceuticals play an important role in personalized nutrition and preventive healthcare strategies.

2.1.3 Functional foods, dietary supplements, and nutraceuticals

The terms functional foods, dietary supplements, and nutraceuticals are often used interchangeably; however, they represent distinct categories within nutrition science. Functional foods are conventional foods that provide health benefits beyond basic nutrition when consumed as part of a regular diet. These foods remain in their natural food form and are not presented as capsules or tablets. Examples include probiotic yogurt, whole grains rich in dietary fiber, and foods fortified with bioactive components such as omega-3 fatty acids.

Dietary supplements are concentrated sources of nutrients or bioactive compounds intended to supplement the diet. They are commonly available in the form of tablets, capsules, powders, or liquids and are used to address specific nutritional deficiencies or to support physiological functions. Nutraceuticals represent a broader umbrella term that encompasses both functional foods and dietary supplements, provided they demonstrate scientifically supported health benefits. Thus, nutraceuticals form a conceptual bridge between food and medicine, emphasizing the role of diet in health promotion and disease prevention (Bagchi and Preuss, 2013).

From a food science and nutrition perspective, nutraceuticals reflect the transition from *nutrition for survival to nutrition for optimal health*. Understanding their definition, scope, and relationship with functional foods and dietary supplements is essential to appreciate how food components interact with human physiology and contribute to long-term health outcomes.

2.2 Historical and conceptual background

The concept of nutraceuticals is deeply rooted in human history, long preceding the development of modern nutrition science and pharmaceutical medicine. Across civilizations, food has been regarded not merely as a source of sustenance but also as a means to maintain health, prevent disease, and restore bodily balance. Traditional dietary practices and medical systems laid the conceptual foundation for what is now scientifically recognized as nutraceuticals.

2.2.1 Ancient use of nutraceuticals

Ancient civilizations intuitively recognized the health-promoting and therapeutic properties of foods. Historical records from Egypt, Mesopotamia, India, and China indicate the use of specific foods and natural substances for strengthening the body, improving digestion, and treating ailments. Foods such as garlic, turmeric, ginger, honey, fermented milk, and herbs were valued not only for their nutritional content but also for their medicinal effects.

In many early societies, there was no clear distinction between food and medicine. Natural foods were selected and consumed based on their observed effects on health and vitality. These empirical practices formed the earliest examples of nutraceutical use, where dietary components were intentionally used to support physiological functions and enhance resistance to disease (Watson, 2015). The continued relevance of these ancient practices highlights the enduring relationship between diet and health, which modern nutraceutical science seeks to validate through systematic research.

2.2.2 Role of nutraceuticals in traditional medicine systems

Traditional medical systems such as Ayurveda, Traditional Chinese Medicine (TCM), and Unani medicine place strong emphasis on diet and natural substances as central elements of health maintenance and disease prevention. In Ayurveda, food is considered a fundamental pillar of health (Ahara), and dietary components are classified according to their effects on body constitution (doshas) and metabolic balance. Herbs, spices, and functional foods such as turmeric, amla, and ghee are traditionally used to enhance immunity, digestion, and overall vitality. These practices align closely with the modern nutraceutical concept of using food-derived bioactive compounds for health promotion (Bagchi and Preuss, 2013).

Traditional Chinese Medicine adopts a holistic approach in which foods are categorized based on their functional properties, such as warming, cooling, moistening, or drying effects. Food therapy is used to regulate energy balance (Qi), improve organ function, and prevent illness. Many plant-based ingredients used in TCM have since been identified as rich sources of bioactive compounds with antioxidant and antiinflammatory properties. In Unani medicine, diet plays a critical role in maintaining humoral balance within the body. Foods and natural substances are prescribed based on their temperament and physiological effects, reinforcing the idea that nutrition and medicine are closely interconnected. These traditional systems demonstrate that the nutraceutical concept is not new but represents a scientific continuation of long-established dietary wisdom.

2.2.3 *Emergence of the modern nutraceutical concept*

The modern concept of nutraceuticals emerged during the late 20th century as scientific research began to establish clear links between diet and the prevention of chronic diseases. Epidemiological studies revealed associations between consumption of certain foods and reduced risk of cardiovascular diseases, cancer, diabetes, and neurodegenerative disorders. This led to increased interest in identifying specific bioactive compounds responsible for these protective effects.

Advances in food chemistry, nutritional biochemistry, and molecular biology enabled researchers to isolate and characterize phytochemicals, bioactive peptides, fatty acids, and microbial metabolites present in foods. These discoveries transformed traditional dietary observations into evidence-based nutritional strategies (Wildman, 2007).

The term nutraceutical gained prominence as a means to describe food-derived substances that offer physiological benefits beyond basic nutrition. This marked a conceptual shift from treatment-focused healthcare to preventive and wellness-oriented nutrition, emphasizing long-term dietary interventions to support health and reduce disease risk (Ross et al., 2020). Today, nutraceuticals represent a scientifically grounded field that integrates traditional knowledge with modern research methodologies, supporting their role in preventive nutrition and public health.

2.3 Scientific classification of nutraceuticals

Nutraceuticals constitute a diverse group of food-derived substances with varying origins, chemical compositions, and physiological effects. To understand their complexity and functional relevance, systematic classification is essential. Scientific classification not only aids in academic understanding but also supports product development, regulatory evaluation, and evidence-based application in health promotion.

Nutraceuticals are commonly classified based on source, chemical nature, and mechanism of action. This multidimensional classification provides a comprehensive framework to evaluate their origin, functional components, and biological effects (Wildman, 2007) (Fig. 2.1).

2.3.1 *Classification based on source*

Classification based on source focuses on the biological origin of nutraceuticals. Different sources contribute unique bioactive compounds, influencing their functionality, bioavailability, and application.

2.3.1.1 *Plant-based nutraceuticals*

Plant-based nutraceuticals represent the largest and most extensively studied category. They are derived from fruits, vegetables, cereals, legumes, oilseeds, herbs, and spices. Plants synthesize a wide range of secondary metabolites that serve protective functions and, when consumed, provide health benefits to humans.

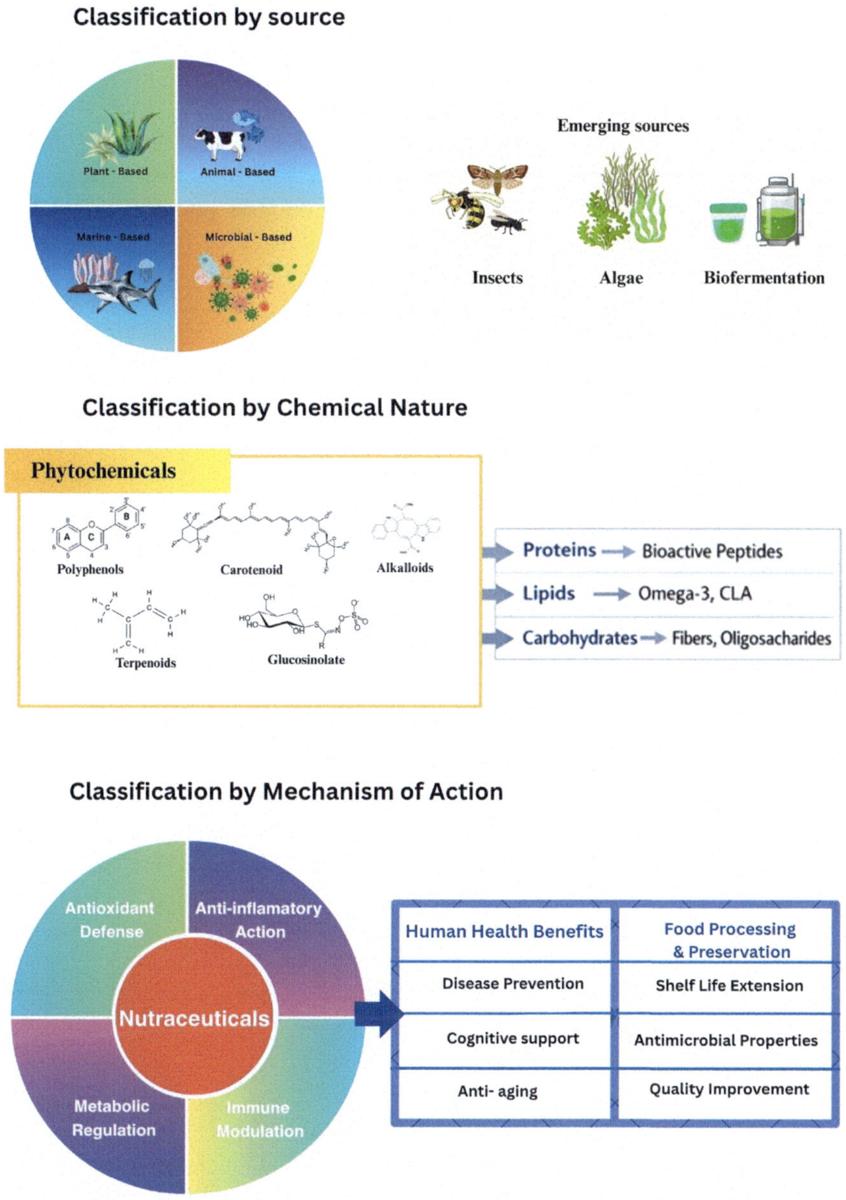

Figure 2.1 Comprehensive classification and benefits of nutraceutical. This image illustrates the diverse sources, chemical nature, and mechanisms of action of nutraceuticals, highlighting their role in promoting human health and enhancing food preservation.

Key plant-based nutraceuticals include polyphenols (flavonoids, phenolic acids), carotenoids (β-carotene, lycopene, lutein), phytosterols, dietary fibers, and sulfur-containing compounds. These compounds exhibit antioxidant, antiinflammatory, cardioprotective, and anticancer properties. Epidemiological studies consistently associate diets rich in plant-based foods with reduced risk of chronic diseases, supporting the nutraceutical potential of plant-derived bioactives (Ross et al., 2020). Plant-based nutraceuticals are widely incorporated into functional foods, dietary supplements, and herbal formulations due to their natural origin and consumer acceptance.

2.3.1.2 Animal-based nutraceuticals

Animal-based nutraceuticals are derived from milk, eggs, meat, and other animal products. These nutraceuticals primarily include bioactive peptides, essential fatty acids, and fat-soluble vitamins. Important examples include omega-3 fatty acids (EPA and DHA) from fish and dairy, bioactive peptides derived from milk proteins, conjugated linoleic acid (CLA), and probiotics from fermented dairy products.

Animal-based nutraceuticals play a crucial role in supporting cardiovascular health, muscle function, immune modulation, and gut health. Milk-derived peptides, for instance, exhibit antihypertensive and antimicrobial activities, highlighting the functional significance of animal-origin nutraceuticals (Bagchi and Preuss, 2013).

2.3.1.3 Marine-based nutraceuticals

Marine organisms are rich sources of nutraceutical compounds with unique structural and functional properties. Fish, algae, seaweed, and marine microorganisms contribute bioactives that are not commonly found in terrestrial foods. Marine-based nutraceuticals include omega-3 fatty acids (EPA, DHA), marine peptides, polysaccharides (alginate, carrageenan), minerals such as iodine and calcium.

These compounds are associated with antiinflammatory, cardioprotective, neuroprotective, and antioxidant effects. Marine nutraceuticals are increasingly valued due to their high bioavailability and therapeutic potential (Watson, 2015).

2.3.1.4 Microbial-based nutraceuticals

Microbial-based nutraceuticals originate from beneficial microorganisms and fermentation processes. This category includes probiotics, prebiotics, and microbial metabolites that influence gut health and immune function. Examples include probiotic bacteria (*Lactobacillus*, *Bifidobacterium*), prebiotic oligosaccharides, organic acids and short-chain fatty acids, and bacteriocins.

These nutraceuticals support intestinal health by modulating gut microbiota, enhancing nutrient absorption, and strengthening host defense mechanisms. The growing understanding of the gut health axis has significantly increased interest in microbial nutraceuticals (Ross et al., 2020).

2.3.1.5 Novel and emerging sources

Recent advances in biotechnology and sustainability research have led to the exploration of novel nutraceutical sources. These include insects, algae, genetically engineered organisms, and biofermentation-derived products. Novel sources are being investigated for their high protein content, bioactive peptides, essential fatty acids, and micronutrients. Although promising, these nutraceuticals require thorough evaluation for safety, acceptability, and regulatory approval before widespread application (Wildman, 2007).

2.3.2 Classification of nutraceuticals based on chemical nature

Classification based on chemical nature focuses on the molecular composition of nutraceuticals and the type of bioactive compounds responsible for their physiological effects. This approach is particularly useful in understanding mechanisms of action, bioavailability, and stability during processing and storage. Nutraceuticals may be broadly categorized into phytochemicals, macronutrient-based compounds, micronutrients, and microbial metabolites (Wildman, 2007; Watson, 2015).

Classification by chemical nature emphasizes that nutraceuticals are not limited to a single group of compounds but encompass a wide range of bioactive molecules. Understanding their chemical diversity provides insight into their functional properties, interactions within the food matrix, and physiological effects.

2.3.2.1 Phytochemicals

Phytochemicals are naturally occurring bioactive compounds synthesized by plants as part of their secondary metabolism. Although not considered essential nutrients, phytochemicals exert significant biological effects that contribute to disease prevention and health promotion. Major classes of phytochemicals include polyphenols (flavonoids, phenolic acids), carotenoids (β-carotene, lycopene, lutein), alkaloids, terpenoids, glucosinolates, and sulfur-containing compounds.

Phytochemicals are widely recognized for their antioxidant, antiinflammatory, anticancer, and cardioprotective properties. Their ability to neutralize free radicals, modulate enzyme activity, and regulate gene expression underlies many of their health benefits. Diets rich in fruits, vegetables, and whole grains are consistently associated with reduced incidence of chronic diseases, highlighting the nutraceutical relevance of phytochemicals (Ross et al., 2020).

2.3.2.2 Macronutrient-based compounds

Macronutrient-based nutraceuticals are derived from proteins, lipids, and carbohydrates, which are traditionally viewed as sources of energy and structural components. However, specific fractions of these macronutrients exhibit bioactive properties that extend beyond basic nutrition. Protein-derived bioactive peptides are released during digestion or food processing and can exert antihypertensive, antioxidant, immunomodulatory, and antimicrobial effects. Milk, egg, and fish proteins are important sources of such peptides. Lipid-based nutraceuticals include essential fatty acids such as omega-3 and omega-6 fatty acids, conjugated linoleic acid, and phospholipids. These compounds play critical roles in cardiovascular health, inflammation regulation, and neural development. Carbohydrate-based nutraceuticals include dietary fibers, resistant starch, and oligosaccharides, which support gastrointestinal health, glycemic control, and cholesterol reduction. These compounds also serve as prebiotics, selectively stimulating beneficial gut microorganisms (Bagchi and Preuss, 2013).

2.3.2.3 Micronutrients

Micronutrients such as vitamins and minerals are essential for normal physiological functioning and are classified as nutraceuticals when used to promote health and reduce disease risk beyond the prevention of deficiency. Examples include vitamins with antioxidant or regulatory roles (vitamins C, E, D) and minerals involved in enzyme function and immune support (zinc, selenium, iron). When provided in optimized forms or fortified foods, micronutrients contribute to immune function, bone health, metabolic regulation, and oxidative balance. Their nutraceutical value lies in their ability to support physiological processes and enhance overall health when consumed at appropriate levels (Ross et al., 2020).

2.3.2.4 Microbial metabolites

Microbial metabolites are bioactive compounds produced during fermentation by beneficial microorganisms. These include organic acids, short-chain fatty acids, bacteriocins, and vitamins synthesized by gut microbiota. Such metabolites play an important role in maintaining gut health, enhancing nutrient absorption, and inhibiting pathogenic microorganisms. Fermented foods rich in microbial metabolites have been associated with improved digestive health and immune function, reinforcing their classification as nutraceutical components (Watson, 2015). In addition, several microbial metabolites exhibit antioxidant, antiinflammatory, and immunomodulatory activities, contributing to the prevention of metabolic and gastrointestinal disorders. Their growing recognition in functional food research highlights the importance of controlled fermentation processes in enhancing the nutraceutical value of foods.

2.3.3 Classification of nutraceuticals based on mechanism of action

Classification based on mechanism of action focuses on how nutraceuticals exert their physiological effects within the human body. Unlike source- or chemistry-based classifications, this approach emphasizes biological function, making it particularly relevant for understanding health outcomes, preventive nutrition, and functional applications in food systems.

Based on their predominant physiological roles, nutraceuticals can be broadly categorized into those involved in disease prevention and health promotion and those contributing to food processing and preservation (Ross et al., 2020).

2.3.3.1 Disease prevention and health promotion

One of the most significant roles of nutraceuticals is their contribution to the prevention of chronic diseases and the promotion of overall health. Many nutraceuticals exert their effects by modulating key biological pathways involved in oxidative stress, inflammation, lipid metabolism, glucose regulation, and immune function. Antioxidant activity is one of the primary mechanisms through which nutraceuticals protect against cellular damage. Phytochemicals such as polyphenols and carotenoids neutralize free radicals, thereby reducing oxidative stress, which is implicated in aging and the development of chronic diseases such as cardiovascular disorders and cancer (Ross et al., 2020).Anti-inflammatory effects are another important mechanism. Bioactive compounds including omega-3 fatty acids, flavonoids, and certain peptides modulate inflammatory pathways by influencing cytokine production and enzyme activity. This contributes to reduced risk of inflammatory conditions and metabolic disorders.

Metabolic regulation is achieved through nutraceuticals that influence lipid and glucose metabolism. Dietary fibers, plant sterols, and bioactive peptides help regulate cholesterol levels, improve insulin sensitivity, and support weight management. These effects are particularly relevant in the prevention of lifestyle-related diseases such as diabetes and obesity (Bagchi and Preuss, 2013).

2.3.3.2 Role in food processing and preservation

In addition to their physiological effects, certain nutraceuticals play a functional role in food processing and preservation. This dual role highlights the intersection between nutraceutical science and food technology. Antioxidant nutraceuticals, such as tocopherols, polyphenols, and carotenoids, help prevent oxidative deterioration of fats and oils, thereby extending shelf life and preserving sensory quality. These compounds protect food products from rancidity and nutrient loss during storage and processing (Wildman, 2007). Antimicrobial nutraceuticals, including organic acids, bacteriocins, and certain plant extracts, inhibit the growth of spoilage and pathogenic microorganisms. Their use supports food safety while reducing reliance on synthetic preservatives. Gut health–promoting nutraceuticals, such as probiotics and prebiotics, are increasingly incorporated into functional foods, linking food processing with health promotion. Their stability during processing and storage is an important consideration in product development. Thus, nutraceuticals contribute not only to consumer health but also to food quality, safety, and shelf stability, reinforcing their multifunctional significance in modern food systems. Immune modulation is also a key mechanism. Probiotics, prebiotics, vitamins, and minerals enhance immune response by supporting gut microbiota balance, improving barrier function, and modulating immune cell activity. Regular consumption of such nutraceuticals contributes to improved resistance to infections and overall immune health (Watson, 2015).

Understanding nutraceuticals through the lens of the mechanism of action provides a holistic view of their benefits. Many nutraceuticals act through multiple mechanisms simultaneously, such as antioxidant protection combined with antiinflammatory or metabolic effects. This multifaceted action highlights the importance of whole-food approaches and balanced dietary patterns in nutraceutical applications.

2.4 Conclusion

Nutraceuticals represent a scientifically validated extension of the long-established concept that food can serve functions beyond basic nourishment. As discussed in this chapter, nutraceuticals occupy a unique position at the interface of food and medicine, combining nutritional value with physiological benefits that contribute to health promotion and disease risk reduction. Their relevance has grown in response to increasing prevalence of lifestyle-related disorders and a global shift toward preventive and wellness-oriented nutrition.

The historical perspective highlights that the foundation of nutraceuticals lies in traditional dietary practices and medical systems such as Ayurveda, Traditional Chinese Medicine, and Unani medicine, where foods and natural substances were integral to maintaining health and restoring balance. Modern nutraceutical science builds upon this traditional knowledge base, transforming empirical observations into evidence-based applications through advances in food chemistry, nutritional biochemistry, and molecular biology.

Scientific classification of nutraceuticals based on source, chemical nature, and mechanism of action provides a structured framework for understanding their diversity and functional significance. Classification by source emphasizes the wide range of plant, animal, marine, microbial, and emerging novel materials that contribute bioactive compounds. Classification by chemical nature highlights the importance of phytochemicals, macronutrient-derived bioactives, micronutrients, and microbial metabolites in influencing physiological processes. Mechanism-based classification further clarifies how nutraceuticals contribute to antioxidant defense, immune modulation, metabolic regulation, and food preservation. Despite their promising potential, nutraceuticals must be approached with scientific rigor. Issues related to bioavailability, standardization, safety, efficacy, and regulatory oversight remain critical considerations. Clear differentiation between food, supplement, and therapeutic claims is essential to ensure consumer safety and maintain scientific credibility.

In conclusion, nutraceuticals represent an important and expanding domain within food and nutrition sciences. A strong understanding of their fundamental concepts, historical roots, and scientific classification is essential for students, researchers, and professionals. As research continues to evolve, nutraceuticals are expected to play an increasingly significant role in preventive nutrition and sustainable health promotion, provided their development and application remain grounded in robust scientific evidence.

AI disclosure

During the preparation of this work the author(s) used Open AI for checking grammatical errors and spell check. After using this tool/service, the author(s) reviewed and edited the content as needed and take(s) full responsibility for the content of the publication.

References

Bagchi, D., Preuss, H.G., 2013. Nutraceuticals and Functional Foods: Regulatory and Health Aspects. CRC Press, Boca Raton.

Ross, A.C., Caballero, B., Cousins, R.J., Tucker, K.L., Ziegler, T.R., 2020. Modern Nutrition in Health and Disease, twelfth ed. Wolters Kluwer, Philadelphia.

Watson, R.R. (Ed.), 2015. Nutraceuticals: Efficacy, Safety and Toxicity. Academic Press, London.

Wildman, R.E.C. (Ed.), 2007. Handbook of Nutraceuticals and Functional Foods, second ed. CRC Press, Boca Raton.

Overview of food additives used in processing and preservation

3

Umalatha[1] *and M.R. Raghunath*[2]
[1]Department of Food Science and Technology, Padmashree Institute of Management and Sciences, Bengaluru, Karnataka, India, [2]Department of Food Science and Technology, Padmashree Institute of Management and Sciences, Bengaluru, Karnataka, India

3.1 Introduction

Food additives have become integral to modern food systems, playing a critical role in ensuring food safety, quality, stability, and consumer acceptability. Unlike conventional ingredients that contribute to bulk or nutritional value, food additives are incorporated in small quantities to perform specific technological functions such as preservation, texture modification, flavor enhancement, color stabilization, and shelf-life extension (Branen et al., 2002; Damodaran et al. 2007). The increasing complexity of food supply chains, urbanization, and the demand for ready-to-eat and processed foods have further intensified the reliance on food additives. Their use allows foods to withstand processing stresses, transportation, and extended storage while maintaining sensory and safety attributes.

3.1.1 Definition of food additives

Food additives are substances intentionally added to foods for technological purposes during manufacturing, processing, packaging, or storage, and are not normally consumed as food.

According to the Codex Alimentarius Commission, a food additive is defined as:

"Any substance not normally consumed as a food by itself and not normally used as a typical ingredient of the food, whether or not it has nutritive value, the intentional addition of which to food for a technological (including organoleptic) purpose result, or may reasonably be expected to result, in it or its by-products becoming a component of, or otherwise affecting the characteristics of such foods" (Codex Alimentarius Commission, 2023).

This definition serves as the global reference framework for food additive regulation and harmonization.

In India, the regulatory authority, FSSAI (Food Safety and Standards Authority of India), adopts a closely aligned definition under the Food Safety and Standards (Food Products Standards and Food Additives) Regulations, 2011, emphasizing the intentional technological role of additives and their use within scientifically established safety limits (FSSAI, 2011).

Nutraceutical Applications in Food Processing and Preservation. DOI: https://doi.org/10.1016/B978-0-443-45784-5.00004-6

Together, these definitions emphasize that food additives are purpose-driven substances, used under regulatory control to improve food quality and safety.

3.1.2 Historical evolution of food additives

The use of substances to preserve and improve foods predates modern food science. Early civilizations relied on natural additives such as salt, vinegar, smoke, and fermentation products to delay spoilage and enhance palatability. These methods function by reducing water activity, lowering pH, or inhibiting microbial growth (Rahman, 2007).

With advances in food chemistry and microbiology during the 20th century, additive use shifted from empirical preservation practices to scientifically designed compounds. Antioxidants were developed to control lipid oxidation, preservatives to inhibit specific spoilage and pathogenic microorganisms, and emulsifiers to stabilize multiphase food systems (Branen et al., 2002).

This transition reflects the evolution of food processing from local, short-term preservation to industrial-scale production requiring uniformity, extended shelf life, and regulatory compliance.

3.1.3 Importance of food additives in modern food systems

In contemporary food systems, additives function as enabling agents that support food safety, quality control, and economic feasibility. Preservatives and antioxidants play a central role in minimizing microbial and chemical spoilage, thereby reducing foodborne risks (Damodaran et al., 2007). In addition, emulsifiers, stabilizers, and thickeners contribute to product consistency and texture, while colorants and flavoring agents enhance consumer appeal. These functions are particularly important in processed and convenience foods, where sensory quality and stability must be maintained over extended storage periods. Thus, food additives operate as functional regulators within food matrices, maintaining desirable attributes that would otherwise deteriorate during processing and storage.

3.1.4 Contemporary relevance and consumer perspective

In recent decades, growing consumer awareness regarding food composition has led to increased scrutiny of food additives, particularly in relation to highly processed foods. This has driven the clean-label movement, which emphasizes simpler ingredient lists and a preference for naturally derived additives. Despite this trend, scientific consensus affirms that approved food additives are safe when used within prescribed limits established through rigorous risk assessment (Codex Alimentarius Commission, 2023).

Consequently, modern food science faces the challenge of balancing technological necessity with consumer expectations. Understanding food additives today, therefore, requires an integrated perspective that considers functionality, safety evaluation, regulatory oversight, and public perception.

3.2 Classification of food additives

Food additives represent a broad and diverse group of substances used in food systems. To understand their role in food processing and preservation, it is essential to classify them systematically. Classification helps in relating the technological function, source, and regulatory identity of additives, thereby providing clarity for food scientists, regulators, and students of food science.

In the food science literature, additives are commonly classified by their functional role in foods and their origin. This dual approach reflects both the technological purpose of additives and their source-related perception and regulatory treatment (Branen et al., 2002; Damodaran et al., 2007).

3.2.1 Classification based on functional role

From a functional perspective, food additives are grouped according to the specific technological role they perform in the food matrix. These roles include preservation, oxidation control, emulsification, stabilization, sweetening, coloring, flavor enhancement, moisture retention, and texture modification. In practical food formulations, multiple additives are often used together, each contributing to a distinct functional requirement.

Preservatives are used to inhibit microbial growth and extend shelf life, while antioxidants delay oxidative deterioration of fats, pigments, and vitamins. Emulsifiers, stabilizers, and thickeners help maintain physical stability and desirable texture in multiphase foods. Sweeteners, colorants, and flavoring agents primarily influence sensory attributes, whereas enzymes, acidity regulators, and anticaking agents support processing efficiency and product consistency (Branen et al., 2002).

Table 3.1 presents a consolidated overview of major functional classes of food additives, along with representative examples, INS numbers, origin, and common food applications.

3.2.2 Classification based on origin

Food additives may also be classified according to their source of origin, which reflects how the additive is obtained or synthesized. Based on origin, food additives are categorized into three major groups:

1. Natural additives—obtained directly from plant, animal, or microbial sources. Example: curcumin, pectin, guar gum, lecithin
2. Nature-identical additives—synthetically produced but chemically identical to naturally occurring compounds. Example: ascorbic acid, vanillin, monosodium glutamate
3. Synthetic additives—entirely artificial compounds not naturally present in foods. Example: tartrazine, BHA, BHT, sucralose

3.3 Functional roles of major food additives

Food additives exert their effects through well-defined physicochemical and biochemical mechanisms within food matrices. While classification provides

Table 3.1 Functional classification of food additives with examples and applications.

Functional class	Representative examples	INS No.	Origin	Typical food applications
Preservatives	Sodium benzoate, potassium sorbate, calcium propionate	211, 202, 282	Synthetic/nature identical	Juices, jams, bakery products
Antioxidants	Ascorbic acid, tocopherols, BHA, BHT	300, 307–309, 320, 321	Natural and synthetic sources	Oils, fats, snacks
Emulsifiers	Lecithin, mono- and diglycerides	322, 471	Natural and synthetic sources	Ice cream, margarine
Stabilizers and thickeners	Xanthan gum, guar gum, carrageenan, pectin	415, 412, 407, 440	Natural (plant/microbial)	Yogurt, sauces, desserts
Sweeteners (nutritive)	Sucrose, glucose, sorbitol	420	Natural/fermentation derived	Bakery, confectionery
Sweeteners (nonnutritive)	Aspartame, sucralose, stevia glycosides	951, 955, 960	Synthetic or plant-derived	Sugar-free foods
Colorants	Tartrazine, curcumin, anthocyanins	102, 100, 163	Synthetic or natural pigments	Beverages, dairy
Flavor enhancers	Monosodium glutamate, disodium inosinate	621, 631	Fermentation-derived/nature-identical	Soups, snacks
Acidulants/pH regulators	Citric acid, lactic acid, sodium citrate	330, 270, 331	Natural/fermentation derived	Beverages, cheese
Anticaking agents	Silicon dioxide, tricalcium phosphate	551, 341	Natural (mineral)	Powdered foods
Humectants	Glycerol, propylene glycol	422, 1520	Synthetic	Confectionery, bakery
Enzymes (processing aids)	Amylase, protease, lipase	—	Natural (microbial/plant/animal)	Baking, brewing
Modified starches	Cross-linked starch, pregelatinized starch	1400–1451	Modified natural	Soups, sauces
Packaging gases	Nitrogen, carbon dioxide, oxygen	941, 290, 948	Natural	MAP systems
Nutrient additives	Vitamins, iron, zinc, calcium	—	Natural or synthetic forms	Fortified foods

Figure 3.1 Functional classes of major food additives. Major categories of food additives used in food processing, including natural preservatives, sweeteners, emulsifiers, acidity regulators, natural gums and gelling agents, natural colors, and flavoring agents.

structural understanding, the functional roles of additives explain how they preserve food quality, safety, and sensory attributes during processing and storage. Among all classes of additives, preservatives and antioxidants occupy a central position due to their direct role in controlling spoilage and deterioration (Fig. 3.1).

3.3.1 Preservatives

Preservatives are substances added to foods to delay spoilage caused by microbial growth or undesirable chemical reactions. Their primary function is to extend shelf life while maintaining safety and quality. Preservatives are broadly classified into antimicrobial preservatives and antioxidant preservatives, based on their mode of action (Branen et al., 2002; Damodaran et al., 2007).

3.3.1.1 Antimicrobial preservatives

Antimicrobial preservatives inhibit the growth of bacteria, yeasts, and molds by interfering with essential cellular processes, including enzyme activity, membrane integrity, and intracellular pH regulation.

Commonly used antimicrobial preservatives include benzoates, sorbates, propionates, nitrites, and sulfur dioxide. Benzoic acid and its salts are particularly effective in acidic foods, where the undissociated form penetrates microbial cell membranes and disrupts intracellular enzyme systems. Sorbates are widely used to control yeasts and molds in bakery products, beverages, and dairy foods due to their broad antifungal activity (Branen et al., 2002). Nitrites, used in cured meat products, play a dual role: they inhibit Clostridium botulinum and contribute to characteristic color and flavor development. However, their use is strictly regulated due to the potential formation of N-nitroso compounds, highlighting the importance of controlled application and regulatory oversight (Damodaran et al., 2007). In essence, antimicrobial preservatives act as protective barriers, preventing microbial proliferation that would otherwise compromise food safety and shelf life.

3.3.1.2 Antioxidant preservatives

Antioxidants are added to foods primarily to retard oxidative deterioration, particularly lipid oxidation, which leads to rancidity, off-flavors, nutrient loss, and discolouration. Oxidation is a free-radical-mediated chain reaction, and antioxidants function by interrupting this process at various stages (Akoh and Min, 2017).

Common food antioxidants include butylated hydroxyanisole (BHA), butylated hydroxytoluene (BHT), tertiary-butylhydroquinone (TBHQ), ascorbic acid, and tocopherols. Synthetic phenolic antioxidants such as BHA and BHT act as hydrogen donors, stabilizing free radicals and terminating oxidative chain reactions in fats and oils (Branen et al., 2002). Natural antioxidants, such as ascorbic acid (vitamin C), tocopherols (vitamin E), carotenoids, and polyphenols (e.g., flavonoids), also play a crucial role in preventing oxidation. They act by donating electrons or hydrogen atoms to neutralize free radicals, chelating pro-oxidant metal ions, and interrupting oxidative chain reactions. These compounds are commonly found in fruits, vegetables, oils, and plant-based foods and are preferred due to their perceived safety and additional health benefits.

3.3.1.2.1 Synergistic action of antioxidants

In food systems, antioxidants are rarely used individually. Instead, combinations of antioxidants are employed to achieve enhanced protective effects through synergism. For example, BHA and BHT are often used together because their combined action provides greater oxidative stability than either compound alone. This synergistic effect arises from differences in solubility, volatility, and radical-scavenging behavior, allowing more comprehensive protection of lipid systems (Akoh and Min, 2017).

3.3.1.2.2 Metal chelators as preservatives

Antioxidant efficacy is further enhanced by adding metal chelating agents such as citric acid and ethylenediaminetetraacetic acid (EDTA). Trace metal ions, iron and copper, catalyze lipid oxidation by initiating free-radical formation. Chelators bind these pro-oxidant metals, thereby reducing their catalytic activity and complementing the action of primary antioxidants (Damodaran et al., 2007).

Thus, antioxidant systems in foods typically consist of:

- Primary antioxidants (e.g., BHA, BHT, tocopherols) that scavenge free radicals
- Secondary antioxidants or synergists (e.g., citric acid, EDTA) that inhibit oxidation promoters

This cooperative strategy significantly improves oxidative stability in fat-rich foods such as edible oils, snacks, cereals, and processed meat products. In practical terms, antioxidants function like protective shields, preserving both the sensory and nutritional quality of foods throughout their intended shelf life.

3.3.2 Emulsifiers, stabilizers, and thickeners

Many processed foods consist of complex, multiphase systems in which components such as oil, water, air, proteins, and carbohydrates coexist. Maintaining physical stability and desirable texture in such systems is a major technological challenge in food processing. Emulsifiers, stabilizers, and thickeners are therefore essential additives that work individually and cooperatively to control structure, consistency, and shelf stability.

Although these three classes are often discussed together, they differ in their primary mode of action within the food matrix. Understanding these differences is critical for rational formulation and quality control (Fig. 3.2).

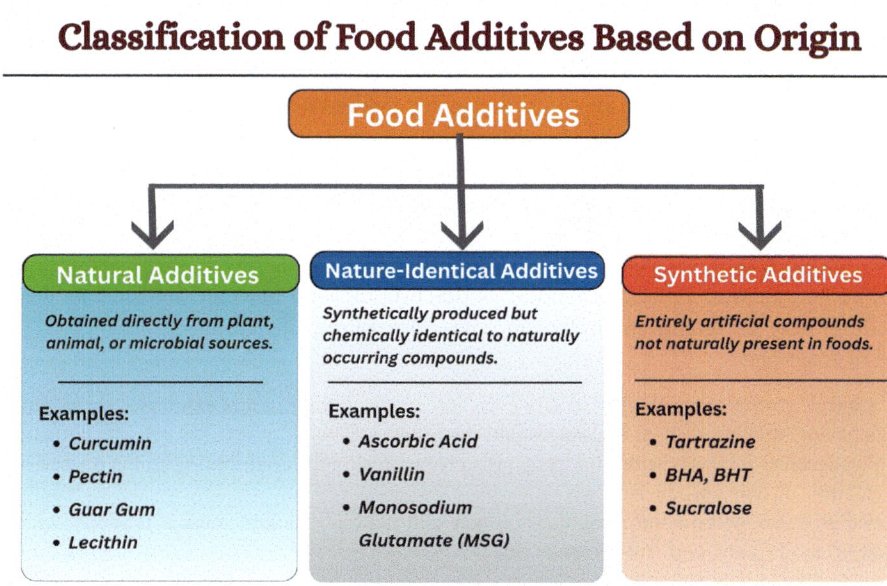

Figure 3.2 Classification of food additives based on origin. Origin-based classification of food additives.

3.3.2.1 Emulsifiers

Emulsifiers are surface-active agents that facilitate the mixing of immiscible phases, most commonly oil and water. They function by reducing interfacial tension and forming a protective layer around dispersed droplets, thereby preventing coalescence and phase separation (McClements, 2015). Chemically, emulsifiers are amphiphilic molecules possessing both hydrophilic (water-attracting) and lipophilic (fat-attracting) regions. This dual affinity allows them to orient at the oil–water interface, stabilizing emulsions.

Common food emulsifiers include lecithin, monoglycerides, diglycerides, and polysorbates. Lecithin, a naturally occurring phospholipid, is widely used in chocolate, bakery products, and emulsified sauces. Mono- and diglycerides are among the most extensively used emulsifiers due to their versatility and compatibility with a wide range of food systems. (Branen et al., 2002).

The effectiveness of an emulsifier depends on its hydrophilic–lipophilic balance (HLB) value, which determines whether it is better suited for oil-in-water or water-in-oil emulsions. Emulsifiers are therefore the primary agents responsible for creating and maintaining dispersed systems such as mayonnaise, ice cream, margarine, and salad dressings.

3.3.2.2 Stabilizers and thickeners

While emulsifiers create and stabilize interfaces, stabilizers and thickeners primarily act by modifying the viscosity and rheological properties of the continuous phase. These additives are typically high-molecular-weight polysaccharides or proteins that interact strongly with water.

Stabilizers function by preventing physical instability such as creaming, sedimentation, or syneresis. They slow down the movement of dispersed particles or droplets by increasing the viscosity of the surrounding medium.

Thickeners, on the other hand, are used mainly to increase consistency and impart body to foods. Although their action overlaps with that of stabilizers, thickeners are often selected specifically to achieve a desired mouthfeel or flow behavior rather than long-term stability alone (Imeson, 2009).

Common stabilizers and thickeners include xanthan gum, guar gum, carrageenan, pectin, and gelatin. These hydrocolloids function through mechanisms such as water binding, polymer chain entanglement, and gel network formation.

For example:

- Xanthan gum provides high viscosity even at low concentrations and exhibits shear-thinning behavior, making it ideal for sauces and dressings.
- Carrageenan interacts with milk proteins to form weak gels, contributing to the texture and stability of dairy products.
- Pectin forms gels in the presence of sugar and acid or calcium ions, a property widely exploited in jams and fruit preparations.

3.3.2.3 Difference in mode of action

Although emulsifiers, stabilizers, and thickeners are often used together, their primary modes of action are distinct:

- Emulsifiers act at interfaces, reducing interfacial tension and stabilizing dispersed droplets.
- Stabilizers act in the bulk phase, preventing phase separation and structural breakdown over time.
- Thickeners primarily increase viscosity and modify texture, influencing flow and mouthfeel.

This distinction is essential for understanding why a single additive cannot replace the combined function of all three classes in complex food systems.

3.3.2.4 Cooperativity in food systems

In practical food formulations, emulsifiers, stabilizers, and thickeners often act cooperatively rather than independently. For example, in ice cream, emulsifiers facilitate fat destablization and air incorporation, whereas stabilizers control ice crystal growth and improve melt resistance. In salad dressings, emulsifiers maintain oil–water dispersion, and thickeners provide viscosity that enhances stability and sensory perception. (McClements, 2015). This cooperative interaction allows formulators to achieve optimal stability, texture, and sensory quality using lower concentrations of individual additives. Such synergistic use also improves processing tolerance and shelf-life performance. Thus, emulsifiers, stabilizers, and thickeners together function as structural architects of food systems, shaping both physical stability and consumer perception.

3.3.3 Sweeteners

Sweeteners are food additives used primarily to impart sweetness, one of the most fundamental sensory attributes influencing food acceptance. Beyond sweetness perception, these additives also affect texture, bulk, browning reactions, mouthfeel, and overall palatability. In modern food systems, sweeteners play an increasingly important role due to rising concerns related to sugar intake, obesity, and metabolic disorders.

Based on their contribution to energy and bulk, sweeteners are broadly classified into nutritive (caloric) sweeteners and nonnutritive (low- or no-calorie) sweeteners (Magnuson et al., 2016; Damodaran et al., 2007).

3.3.3.1 Nutritive sweeteners

Nutritive sweeteners provide sweetness along with energy, typically contributing approximately 4 kcal per gram. Common examples include sucrose, glucose, fructose, and sugar alcohols such as sorbitol. These sweeteners not only enhance sweetness but also contribute to bulk, viscosity, and structure in foods.

Sucrose is considered the reference standard for sweetness and is widely used in bakery products, confectionery, dairy desserts, and beverages. In addition to sweetness, sucrose contributes to color and flavor development through caramelization reactions and, upon hydrolysis into glucose and fructose, indirectly participates in Maillard browning (Damodaran et al., 2007).

Sugar alcohols, such as sorbitol and xylitol, provide sweetness with a lower caloric value than sucrose and are commonly used in sugar-free confectionery and diabetic-friendly foods. However, excessive consumption of certain polyols may cause gastrointestinal discomfort due to incomplete absorption

3.3.3.2 Nonnutritive and low-calorie sweeteners

Nonnutritive sweeteners are compounds that provide intense sweetness at very low concentrations, contributing negligible or no energy to the diet. These sweeteners are particularly important in products formulated for calorie reduction, weight management, and glycemic control.

Common nonnutritive sweeteners include aspartame, sucralose, acesulfame-K, and steviol glycosides. Aspartame is approximately 200 times sweeter than sucrose but is not heat-stable, limiting its use in baked products. Sucralose, which is around 600 times sweeter than sucrose, is heat-stable and therefore suitable for a wider range of applications, including cooking and baking (Magnuson et al., 2016). In addition, these sweeteners provide negligible or no calories and are widely used in low-calorie and diabetic-friendly formulations. However, due to their high sweetness intensity, they are often blended with other sweeteners or bulking agents to improve the taste profile and texture in food products.

Natural high-intensity sweeteners such as steviol glycosides (from *Stevia rebaudiana*) and monk fruit extract (from *Siraitia grosvenorii*) have gained significant attention due to their plant-based origin and suitability for clean-label food formulations. These sweeteners are noncaloric or very low in calories. They exhibit high sweetness intensity compared to sucrose and are generally heat-stable, allowing their use in a variety of processed foods and beverages. While steviol glycosides may impart a slight bitter or liquorice-like aftertaste, monk fruit sweeteners typically provide a cleaner taste profile, although a mild lingering sweetness may be perceived at higher concentrations.

3.3.3.3 Effects on metabolism and glycemic control

The increasing use of nonnutritive sweeteners has been driven by the need to reduce dietary sugar intake, particularly in individuals with diabetes mellitus. Unlike nutritive sweeteners, nonnutritive sweeteners do not cause a rapid rise in blood glucose levels and are therefore useful in glycemic control (Magnuson et al., 2016). However, emerging research suggests that while nonnutritive sweeteners are effective in reducing caloric intake, their long-term effects on metabolism, appetite regulation, and gut microbiota require careful consideration. Current scientific consensus supports their safety within established acceptable daily intake (ADI) levels, but emphasizes moderation and context-specific use (Damodaran et al., 2007). Thus, sweetener selection in food formulation must balance sweetness intensity, metabolic impact, sensory quality, and regulatory compliance.

3.3.3.4 Functional importance in food systems

From a technological perspective, sweeteners influence more than sweetness alone. Nutritive sweeteners contribute to bulk, texture, freezing point depression, and shelf stability, while nonnutritive sweeteners often require bulking agents or texture modifiers to compensate for the loss of mass and mouthfeel. In modern food systems, sweeteners therefore function as multifunctional additives, affecting sensory perception, nutritional profile, and physical properties of foods. Their appropriate selection and combination are essential for developing products that meet both health objectives and consumer expectations.

3.3.4 Colorants

Color is one of the most influential sensory attributes of food, often shaping consumer perception even before aroma or taste is evaluated. Colorants are therefore used to restore color lost during processing, enhance natural appearance, or provide uniformity across batches. In many foods, color serves as a visual indicator of freshness, flavor intensity, and quality, making color control a critical aspect of product formulation (Damodaran et al., 2007). Food colorants are broadly classified into synthetic and natural colorants based on their origin, each category offering distinct technological advantages and limitations.

3.3.4.1 Synthetic colourants

Synthetic colorants are chemically synthesized compounds characterized by high coloring strength, stability, and consistency. Common examples include tartrazine, sunset yellow FCF, Allura red AC, and brilliant blue FCF. These colorants are widely used in beverages, confectionery, bakery products, and desserts due to their excellent solubility, heat stability, and resistance to light and pH variations (Scotter, 2005).

From a technological perspective, synthetic colorants offer several advantages:

- High tinctorial strength, allowing effective coloring at very low concentrations
- Consistent shade and reproducibility across batches
- Good stability during processing and storage

However, synthetic colorants have been the subject of regulatory scrutiny and public concern due to reports of hypersensitivity reactions and behavioral effects in sensitive populations. As a result, their use is strictly regulated, with clearly defined acceptable daily intake (ADI) limits established through toxicological evaluation (Codex Alimentarius Commission, 2023).

3.3.4.2 Natural colourants

Natural colorants are derived from plant, animal, or microbial sources and include pigments such as curcumin, anthocyanins, carotenoids, chlorophylls, and beetroot red. These colorants are increasingly preferred due to consumer demand for naturally derived ingredients and clean-label products (Damodaran et al., 2007). Natural colorants often offer additional benefits beyond coloring, such as antioxidant activity or bioactive potential. For example, curcumin exhibits antioxidant properties, while anthocyanins are associated with polyphenolic health benefits.

Despite these advantages, natural colorants present certain technological challenges:

- Lower stability to heat, light, and pH changes
- Greater batch-to-batch variability
- Higher cost compared to synthetic alternatives

Advances in encapsulation and formulation strategies have improved the stability and applicability of natural pigments, expanding their use in beverages, dairy products, and nutraceutical formulations (Scotter, 2005).

3.3.4.3 Comparison of natural and synthetic colourants

From a functional standpoint, the choice between natural and synthetic colorants depends on the intended application, processing conditions, regulatory requirements, and consumer expectations.

- Synthetic colorants offer superior stability, cost-effectiveness, and color intensity but face greater consumer resistance.
- Natural colorants align with clean-label trends and consumer perception of healthfulness but require careful formulation to address stability issues.

Thus, colorants serve not merely as esthetic enhancers, but as strategic formulation tools that balance technological performance with regulatory compliance and consumer acceptance.

3.3.4.4 Regulatory considerations

The use of food colorants is subject to strict regulatory control. Internationally, colorants are assigned INS numbers and permitted levels under the General Standard for Food Additives established by the Codex Alimentarius Commission. National authorities, including the Food Safety and Standards Authority of India, specify approved colorants and usage limits for different food categories. Clear labeling requirements ensure transparency and allow consumers to make informed choices, particularly for additives associated with sensitivity concerns.

3.3.5 Flavors and flavor enhancers

Flavoring agents, often referred to as added flavors, are substances incorporated into foods to impart or modify characteristic aromas and tastes that may be lost during processing or storage. These flavoring substances may be derived from natural sources such as spices, herbs, fruits, and essential oils, or produced as nature-identical or synthetic compounds that replicate the chemical structures responsible for specific flavor notes. In processed foods, flavoring agents are widely used to restore the original sensory profile of raw ingredients, enhance product identity, and maintain consistency across batches. Common examples include vanillin for vanilla flavor, citrus oils for beverages, and various esters and aldehydes that contribute fruity or sweet aromas. By directly contributing recognizable flavor characteristics, added flavors play a crucial role in maintaining the sensory quality and consumer acceptability of many processed foods.

3.3.5.1 Major flavor enhancers and their mode of action

The most widely used flavor enhancer is monosodium glutamate (MSG), the sodium salt of the amino acid glutamic acid. Glutamate is naturally present in many foods such as tomatoes, cheese, mushrooms, and fermented products. When added to food, MSG enhances savory taste by eliciting the umami sensation, a distinct basic taste associated with glutamate (Yamaguchi and Ninomiya, 2000).

Other important flavor enhancers include:

- Disodium inosinate (IMP)
- Disodium guanylate (GMP)

These nucleotide-based compounds exhibit strong synergistic effects when used in combination with glutamate, resulting in a much greater enhancement of umami perception than when used individually. This synergism allows effective flavor enhancement at low concentrations, making them particularly valuable in low-sodium and reduced-fat foods. Thus, flavor enhancers act by increasing taste signal intensity at the receptor level rather than altering the chemical composition of the food.

3.3.5.2 Applications in food systems

Flavor enhancers are widely used in soups, sauces, snack foods, seasoning mixes, instant noodles, and processed meat products. Their use is especially important in reformulated foods where reductions in salt, sugar, or fat may otherwise compromise sensory quality. In such applications, flavor enhancers help maintain consumer acceptability while supporting nutritional reformulation goals. For example, umami enhancement can partially compensate for reduced sodium levels, contributing to flavor balance without excessive salt addition (Yamaguchi and Ninomiya, 2000).

3.3.5.3 Natural sources and clean-label alternatives

Although MSG and nucleotide enhancers are often industrially produced, their functional components are identical to naturally occurring compounds. In response to clean-label demands, natural sources of umami such as yeast extract, hydrolyzed vegetable protein, mushroom extracts, seaweed (kombu), and fermented ingredients are increasingly used as alternatives.

These ingredients contain naturally occurring glutamates and nucleotides and function similarly to purified flavor enhancers, though their composition may be less standardized. From a technological standpoint, they offer flavor enhancement while aligning with consumer preference for recognizable ingredients (Branen et al., 2002).

3.3.5.4 Safety and regulatory considerations

Flavor enhancers such as MSG are among the most extensively studied food additives. Scientific evaluations have consistently concluded that MSG is safe for the general population when consumed at customary dietary levels. Regulatory authorities have established acceptable daily intake values or have determined that intake levels from food use do not pose a safety concern (Walker and Lupien, 2000). Despite persistent public misconceptions, controlled clinical studies have not demonstrated consistent or reproducible adverse effects attributable to MSG in the general population. Nevertheless, transparent labeling and adherence to regulatory limits remain essential to maintain consumer trust.

3.3.5.5 Functional significance

From a functional perspective, flavor enhancers play a critical role in optimizing sensory quality, improving flavor balance, and supporting product reformulation strategies. They enable food manufacturers to deliver palatable foods with reduced levels of salt or fat while maintaining consumer acceptance. In this sense, flavor enhancers function as sensory amplifiers, strengthening existing flavor signals and enhancing the overall eating experience without fundamentally altering the food's composition.

3.3.6 Texture modifiers

Texture is a critical quality attribute that strongly influences consumer perception, acceptability, and overall eating experience. Attributes such as thickness, creaminess, firmness, chewiness, spreadability, and mouth-coating are largely determined by the physical structure of the food matrix. Texture modifiers are therefore used to design and control the rheological and structural properties of foods, ensuring consistency during processing, storage, and consumption (Damodaran et al. 2007). Texture modifiers include a wide range of substances such as hydrocolloids, proteins, and modified starches, which interact primarily with water and other food components to create structured systems.

3.3.6.1 Mechanisms of texture modification

Texture modifiers function through several physicochemical mechanisms:

- Water binding and immobilization, which increases viscosity and reduces syneresis
- Polymer chain entanglement, leading to thickening and viscoelastic behavior
- Gel network formation, producing structured, solid-like textures

These mechanisms depend on the molecular size, charge, and conformation of the texture-modifying agent, as well as environmental factors such as pH, temperature, and ionic strength (Phillips and Williams, 2009).

3.3.6.2 Common texture modifiers and their functional roles

Hydrocolloids are among the most widely used texture modifiers in food systems. Examples include xanthan gum, guar gum, gelatin, carrageenan, pectin, and modified starches.

- Xanthan gum produces high viscosity at low concentrations and exhibits shear-thinning behavior, making it ideal for sauces, dressings, and beverages that require ease of pouring combined with stability.
- Guar gum is a galactomannan that hydrates rapidly and enhances thickness and mouthfeel in dairy products and bakery formulations.
- Gelatin, derived from collagen, forms thermoreversible gels that melt at body temperature, contributing to the characteristic texture of jellies and desserts.
- Pectin forms gels through sugar-acid or calcium-mediated mechanisms and is essential in jams, fruit preparations, and low-fat spreads.

- Modified starches provide body, stability, and freeze–thaw resistance in soups, sauces, and ready-to-eat foods (Imeson, 2009).

These additives are selected based on the desired textural outcome and processing conditions.

3.3.6.3 Role in process stability and shelf life

In addition to sensory attributes, texture modifiers contribute significantly to process stability and shelf life. By binding water and stabilizing structural networks, they help prevent moisture migration, phase separation, and textural degradation during storage. For example, in frozen desserts, texture modifiers limit ice crystal growth, improving smoothness and melt resistance. In low-fat formulations, they compensate for the loss of fat-derived mouthfeel, allowing the development of acceptable reduced-calorie products (Damodaran et al., 2007).

3.3.6.4 Interaction with other additives

Texture modifiers often act in coordination with emulsifiers and stabilizers. While emulsifiers manage interfacial stability, texture modifiers build the bulk structure that supports the dispersed system. Such interactions are essential in complex foods such as ice cream, yoghurt, sauces, and bakery fillings. The combined use of different texture modifiers allows fine-tuning of rheological properties and sensory perception while minimizing the concentration of individual additives, thereby enhancing both functionality and consumer acceptability (Phillips and Williams, 2009).

3.3.6.5 Functional significance

From a functional perspective, texture modifiers serve as structural designers of food systems. They translate formulation intent into tangible sensory experiences, ensuring that foods meet consumer expectations for consistency, stability, and mouthfeel. Thus, texture modifiers are indispensable in modern food formulation, enabling the development of products that are not only stable and safe but also pleasurable to consume.

3.4 Regulatory framework and safety evaluation

The use of food additives is governed by comprehensive regulatory frameworks designed to ensure that these substances are safe for consumption and technologically justified. Unlike conventional food ingredients, additives are subject to rigorous scientific evaluation because they are intentionally added to foods for specific functional purposes and may be consumed regularly over long periods. Regulatory oversight, therefore, plays a critical role in protecting public health while enabling technological innovation in food processing. At the international and national levels, food additive regulations are based on toxicological assessment, exposure evaluation, and risk–benefit analysis, ensuring that approved additives are safe when used under specified conditions.

3.4.1 International regulatory framework

3.4.1.1 Codex Alimentarius Commission

At the global level, food additive regulation is harmonized through the Codex Alimentarius Commission, established jointly by the Food and Agriculture Organisation (FAO) and the World Health Organisation (WHO). The Codex system provides internationally recognized standards, guidelines, and codes of practice that serve as reference points for national food laws and international trade. The General Standard for Food Additives (GSFA) classifies food additives according to functional class, assigns International Numbering System (INS) codes, and specifies permitted uses and maximum levels in different food categories (Codex Alimentarius Commission, 2023). The GSFA ensures consistency and transparency in the regulation of additives across countries.

3.4.1.2 Safety evaluation by JECFA

Scientific evaluation of food additives at the international level is conducted by the Joint FAO/WHO Expert Committee on Food Additives (JECFA). JECFA assesses toxicological data from animal studies, human studies, and metabolic investigations to determine the safety of food additives.

A key outcome of JECFA evaluation is the establishment of:

- NOAEL (no observed adverse effect level)—the highest dose at which no adverse effects are observed
- ADI (acceptable daily intake)—the amount of an additive that can be consumed daily over a lifetime without appreciable health risk

These values form the scientific basis for regulatory limits adopted by Codex and national authorities (WHO and FAO, 1987).

3.4.2 National regulatory authorities

3.4.2.1 Food Safety and Standards Authority of India (FSSAI)

In India, food additives are regulated by the Food Safety and Standards Authority of India under the Food Safety and Standards Act, 2006. The Food Safety and Standards (Food Products Standards and Food Additives) Regulations, 2011, specify:

- Approved food additives and their INS numbers
- Permitted food categories
- Maximum allowable levels
- Labeling requirements

The Indian regulatory framework closely aligns with Codex standards while addressing country-specific dietary patterns and food categories (FSSAI, 2011).

3.4.2.2 Other international regulatory bodies

Other major regulatory authorities include the European Food Safety Authority (EFSA) and the US Food and Drug Administration (FDA). EFSA periodically re-evaluates approved additives using updated scientific data, while the FDA regulates additives through the food additive petition process or the Generally Recognized as Safe (GRAS) framework (USFDA, 2023).

Although regulatory approaches differ, all systems rely on scientific risk assessment and continuous monitoring to ensure consumer safety.

3.4.3 Labeling and consumer information

Regulations universally require that food additives be clearly declared on product labels. Typically, additives must be listed by:

- Functional class (e.g., preservative, colorant)
- Specific name or INS number

For example, preservatives such as sodium benzoate (INS 211) and color additives such as curcumin (INS 100).

Certain additives, such as sulfites, require additional warning statements when present above specified levels due to potential sensitivity in susceptible individuals. Transparent labeling enables informed consumer choice and builds trust in food systems (Codex Alimentarius Commission, 2023).

3.4.4 Public concerns and ongoing reevaluation

Despite regulatory approval, some food additives continue to attract public concern, particularly synthetic colorants, nitrites, and sulfites. These concerns have prompted periodic re-evaluation of additives using updated toxicological and exposure data.

Regulatory authorities emphasize that safety evaluation of food additives is an ongoing process rather than a one-time assessment. Approved additives are subject to continuous reevaluation, and their use may be restricted or withdrawn if new scientific evidence indicates potential risk. This adaptive approach ensures that food additive regulation remains responsive to advances in scientific knowledge (EFSA, 2021).

3.4.5 Significance of regulatory oversight

Effective regulation of food additives ensures that:

- Only scientifically evaluated additives are permitted
- Additives are used at technologically justified levels
- Consumer health is protected from long-term exposure

Thus, regulatory frameworks serve as the cornerstone of safe additive use, balancing technological necessity with public health protection.

3.5 Emerging trends and challenges

The field of food additives is undergoing a significant transformation driven by changing consumer expectations, advances in food processing technologies, and evolving regulatory perspectives. While additives remain essential for ensuring food safety, stability, and quality, their selection and use are increasingly influenced by demands for transparency, natural ingredients, and minimal processing.

This evolving landscape presents both opportunities and challenges for food scientists and manufacturers, requiring a careful balance between technological functionality and public perception.

3.5.1 Shift toward clean-label and natural additives

One of the most prominent trends in recent years is the growing demand for *clean-label* foods—products formulated with fewer ingredients and additives perceived as natural, familiar, and minimally processed. This trend has led to increased interest in naturally derived additives such as plant extracts, essential oils, natural colorants, and fermentation-derived ingredients.

Natural preservatives and antioxidants, including rosemary extract, green tea polyphenols, tocopherols, and organic acids, are increasingly being explored as alternatives to synthetic additives. These ingredients are often perceived as safer and healthier by consumers, even though their functional role may be similar to that of conventional additives (Baines and Seal, 2012).

However, natural additives often exhibit greater variability in composition and lower stability under processing conditions, necessitating careful formulation and validation to ensure consistent performance.

3.5.2 Multifunctional additives with dual benefits

An important development in additive research is the use of multifunctional additives that provide more than one technological benefit. Such additives improve formulation efficiency while aligning with clean-label expectations.

A notable example is rosemary extract, which functions both as a natural antioxidant and as a flavor-protecting agent. Its phenolic compounds effectively inhibit lipid oxidation, thereby extending shelf life, while also helping to preserve desirable flavor notes in fat-rich foods (Shahidi and Ambigaipalan, 2015).

Similarly, certain plant extracts and fermentation-derived ingredients offer combined antioxidant, antimicrobial, and sensory benefits. These dual-function additives reduce the need for multiple separate additives and support simpler ingredient lists without compromising food quality.

3.5.3 Food safety versus clean-label preference

While the shift toward natural and clean-label additives reflects legitimate consumer concerns, it is essential to emphasize that food safety cannot be compromised in the pursuit of additive reduction. Additives are incorporated based on scientific evidence and regulatory approval to control microbial growth, oxidation, and other forms of deterioration.

Replacing effective preservatives with less robust natural alternatives without adequate validation may increase the risk of spoilage or foodborne illness. Therefore, the selection of additives—whether natural or synthetic—must always be guided by safety, functionality, and regulatory compliance rather than perception alone (Branen et al., 2002).

3.5.4 Alternative preservation technologies reducing additive dependence

In response to clean-label demands and safety considerations, alternative preservation strategies are increasingly being explored to complement or reduce reliance on chemical additives. These technologies focus on controlling spoilage through physical or environmental means rather than additive-based intervention.

Key approaches include:

- Modified atmosphere packaging (MAP): Alters the gaseous environment around food to slow microbial growth and oxidative reactions.
- High-pressure processing (HPP): Inactivates microorganisms while preserving sensory and nutritional quality.
- Advanced packaging systems: Incorporate barrier materials and active packaging elements to enhance shelf life.

Such technologies can, in some cases, obviate the need for certain additives, particularly preservatives, while still maintaining food safety and quality (Rahman, 2007).

However, these methods often involve higher costs and infrastructure requirements, making them more suitable for specific product categories rather than universal application.

3.5.5 Future challenges and research directions

The future of food additives lies in developing solutions that integrate safety, functionality, and consumer acceptance. Key challenges include:

- Ensuring the consistent performance of natural additives
- Understanding interactions among multiple additives in complex food systems
- Improving consumer literacy regarding the safety evaluation of additives
- Integrating additive use with nonthermal and packaging-based preservation strategies

Continued research, transparent regulation, and science-based communication are essential to ensure that food additives continue to serve their intended role without compromising public trust.

3.6 Conclusion

Food additives play an indispensable role in modern food systems by ensuring product safety, stability, sensory quality, and technological efficiency during processing and storage. Through their diverse functional roles—such as preservation, oxidation control, emulsification, texture modification, sweetening, coloring, and flavor enhancement—additives enable the production of foods that meet both industrial requirements and consumer expectations. As discussed in this chapter, systematic classification based on functional roles and origin provides a framework for understanding how these substances interact within complex food matrices and contribute to product quality and shelf life.

Equally important is the regulatory framework governing food additives. International and national authorities, including the Codex Alimentarius Commission, JECFA, and FSSAI, ensure that additives are subjected to rigorous safety evaluations and used only within scientifically established limits. Such regulatory oversight safeguards public health while enabling technological innovation in food processing. Transparent labeling and continuous reassessment of additives further strengthen consumer confidence in the safety of the food supply.

Looking ahead, the field of food additives is evolving in response to growing consumer demand for natural ingredients, clean-label formulations, and minimally processed foods. Advances in food science, biotechnology, and packaging technologies are creating opportunities to develop multifunctional and naturally derived additives that combine safety, functionality, and sustainability. Future research will therefore focus on optimizing additive performance, understanding interactions within complex food systems, and balancing technological necessity with consumer perception to ensure the continued development of safe, high-quality food products.

AI disclosure

During the preparation of this work, the author(s) used Open AI. ChatGPT (OpenAI) was utilized for assistance in refining and organizing the text. After using this tool/service, the author(s) reviewed and edited the content as needed and take(s) full responsibility for the content of the publication.

References

Akoh, C.C., Min, D.B. (Eds.), 2017. Food Lipids: Chemistry, Nutrition, and Biotechnology, 4th ed. CRC Press.

Baines, D., Seal, R., 2012. Natural Food Additives, Ingredients and Flavourings. Elsevier Ltd., United Kingdom, pp. 1–460. Available from: http://www.sciencedirect.com/science/book/9781845698119. Available at 10.1533/9780857095725.

Branen, A.L., Davidson, P.M., Salminen, S., Thorngate, J.H., 2002. Food Additives, second ed. CRC Press, New York. Available at https://doi.org/10.1201/9780203904030.

Codex Alimentarius Commission, 2023. General Standard for Food Additives (GSFA). FAO/WHO, Rome. Available at https://www.fao.org/fao-who-codexalimentarius/codex-texts/list-standards/en/.

Damodaran, S., Parkin, K.L., Fennema, O.R., 2007. Fennema's Food Chemistry, 4th ed. CRC Press. Available at https://doi.org/10.1201/9781420027976.

European Food Safety Authority (EFSA), 2021. Re-evaluation of food additives: principles and procedures. EFSA Journal.

Food Safety and Standards Authority of India (FSSAI). Food Safety and Standards (Food Products Standards and Food Additives) Regulations, 2011. Available at https://www.fssai.gov.in/cms/food-safety-and-standards-regulations.php.

Imeson, A., 2009. Food Stabilisers, Thickeners and Gelling Agents. Wiley-Blackwell, United Kingdom, pp. 1–352. Available from: http://onlinelibrary.wiley.com/book/10.1002/9781444314724. Available at https://doi.org/10.1002/9781444314724.

Magnuson, B.A., Carakostas, M.C., Moore, N.H., Poulos, S.P., Renwick, A.G., 2016. Biological fate of low-calorie sweeteners. Nutrition Reviews 74 (11), 670–689. Available at https://doi.org/10.1093/nutrit/nuw032. Available from: http://nutritionreviews.oxfordjournals.org/.

McClements, D.J., 2015. Food Emulsions: Principles, Practices, and Techniques, third ed. CRC Press, Boca Raton. Available at https://doi.org/10.1201/b18661.

Phillips, G.O., Williams, P.A. (Eds.), 2009. Handbook of Hydrocolloids, second ed. Woodhead Publishing, Cambridge. Available at https://doi.org/10.1533/9781845695873.

Rahman, M.S. (Ed.), 2007. Handbook of Food Preservation, second ed. CRC Press, Boca Raton. Available at https://doi.org/10.1201/9781420017373.

Scotter, M.J., 2005. The chemistry and analysis of food colourants. Food Additives and Contaminants 22 (6), 557–565. Available at https://doi.org/10.1080/02652030500111905.

Shahidi, F., Ambigaipalan, P., 2015. Phenolics and polyphenolics in foods: antioxidant activity and health effects. Journal of Functional Foods 18, 820–897. Available at https://doi.org/10.1016/j.jff.2015.06.018.

U.S. Food and Drug Administration (FDA). *Overview of Food Ingredients, Additives and Colors*. 2023. https://www.fda.gov/food/food-additives-petitions/overview-food-ingredients-additives-colors.

Walker, R., Lupien, J.R., 2000. The safety evaluation of monosodium glutamate. The Journal of Nutrition 130 (4), 1049S. Available at https://doi.org/10.1093/jn/130.4.1049S.

WHO and FAO, 1987. Principles for the Safety Assessment of Food Additives and Contaminants in Food. Environmental Health Criteria 70. Available at https://apps.who.int/iris/handle/10665/37287.

Yamaguchi, S., Ninomiya, K., 2000. Umami and food palatability. The Journal of Nutrition 130 (4), 921S. Available at https://doi.org/10.1093/jn/130.4.921s.

Nutraceutical and natural food additives

4

M. Anuradha[1] *and S. Balasubramanya*[2]
[1]Padmashree Institute of Management and Sciences, Bengaluru, Karnataka, India,
[2]Association for Biotechnology Led Enterprises (ABLE), Bengaluru, Karnataka, India

4.1 Introduction

Food additives have long been employed in food processing to ensure safety, stability, sensory acceptability, and extended shelf life. Conventional additives are primarily selected for their technological efficiency, with limited emphasis on their nutritional or physiological relevance. However, growing scientific evidence linking diet to long-term health outcomes, along with increased consumer demand for clean-label and health-oriented foods, has driven a reassessment of additive selection strategies y.

In this context, nutraceuticals have gained prominence as functional food additives capable of delivering dual benefits technological performance in food systems and physiological advantages to consumers. Unlike traditional additives that serve a single purpose, nutraceutical-based additives integrate food preservation and quality enhancement with the delivery of bioactive compounds associated with disease risk reduction and health maintenance. The incorporation of nutraceuticals as food additives reflects a shift from additive safety alone toward additive functionality with health value. Natural antioxidants, bioactive fibers, plant pigments, and lipid-derived nutraceuticals are increasingly employed to replace or reduce synthetic additives, thereby aligning food formulation with both regulatory expectations and consumer perception.

This chapter critically examines nutraceuticals specifically in their role as food additives, focusing on their functional applications, health relevance, regulatory considerations, and practical challenges in food systems.

4.2 Conceptual basis of nutraceuticals as food additives

Nutraceuticals are commonly defined as food-derived components that provide health benefits beyond basic nutrition. When these compounds are incorporated into foods primarily for technological purposes, such as preservation, color stabilization, texture modification, or flavor enhancement, they function as nutraceutical food additives.

A key conceptual distinction must be made between nutraceuticals used as supplements and those used as food additives. Supplements are typically consumed in concentrated doses for targeted physiological effects, whereas nutraceutical additives are incorporated at lower levels and consumed as part of the habitual diet. This mode of delivery allows continuous exposure to bioactive compounds while simultaneously fulfilling technological requirements in food processing (Shahidi, 2012).

Nutraceutical Applications in Food Processing and Preservation. DOI: https://doi.org/10.1016/B978-0-443-45784-5.00012-5

For example, tocopherols added to edible oils primarily function as antioxidants, delaying lipid oxidation and extending shelf life. At the same time, these compounds contribute to antioxidant defense in the human body by scavenging free radicals and protecting cellular membranes (Akoh & Min, 2017). Similarly, dietary fibers incorporated as stabilizers or thickeners improve food texture while supporting gut health and metabolic regulation upon consumption.

Thus, nutraceutical food additives represent a functional convergence of food technology and nutrition science, transforming additives from inert technological agents into bioactive contributors to health.

4.3 Dual functional role of nutraceutical food additives

4.3.1 Technological functions in food systems

From a technological perspective, nutraceutical additives perform several essential roles required for maintaining food quality and safety. One of the most significant functions is antioxidant activity. Oxidative reactions are a major cause of quality deterioration in foods, leading to rancidity, off-flavors, color loss, and nutrient degradation. Natural antioxidants such as polyphenols, tocopherols, and carotenoids inhibit free-radical chain reactions, thereby improving oxidative stability of fats, oils, and processed foods (Damodaran et al., 2007).

Another critical function is antimicrobial activity. Certain nutraceutical compounds, including essential oils, organic acids, and phenolic extracts, inhibit microbial growth by disrupting cell membranes, interfering with enzyme systems, or altering intracellular pH. For instance, rosemary and green tea extracts reduce microbial spoilage in meat and bakery products while avoiding the use of synthetic preservatives (Branen et al., 2002).

Nutraceutical additives also contribute to color stabilization and enhancement. Natural pigments protect sensitive food components from oxidative degradation while maintaining visual appeal. In addition, dietary fibers and protein-based nutraceuticals act as emulsifiers, stabilizers, and thickeners, improving texture, viscosity, and physical stability of complex food systems (Imeson, 2009).

4.3.2 Health-promoting functions

In addition to their technological roles, nutraceutical food additives deliver bioactive compounds associated with multiple physiological benefits. Natural antioxidants reduce oxidative stress, a key contributor to chronic diseases such as cardiovascular disorders, diabetes, and neurodegenerative conditions (Shahidi & Ambigaipalan, 2015).

Dietary fibers and polysaccharides used as stabilizers exhibit prebiotic effects, promoting beneficial gut microbiota and improving intestinal health. Bioactive lipids such as omega-3 fatty acids contribute to cardiovascular protection, while phytosterols incorporated into foods reduce intestinal cholesterol absorption (Magnuson et al., 2016). For example, replacing synthetic antioxidants in breakfast cereals with polyphenol-rich plant extracts not only improves oxidative stability during storage but also enhances antioxidant intake, illustrating the dual technological and nutritional value of nutraceutical food additives.

4.4 Classification of nutraceutical-based food additives

The major classes of nutraceutical-based food additives, along with their technological roles and health relevance, are summarized in Table 4.1.

4.4.1 *Nutraceutical preservatives*

Nutraceutical preservatives are derived from natural sources and function by inhibiting oxidative and microbial spoilage. Examples include rosemary extract, green tea catechins, grape seed extract, essential oils, and fermentation-derived organic acids. These compounds are rich in phenolics that act as free-radical scavengers and metal chelators, thereby delaying lipid oxidation (Shahidi, 2012). In addition to preservation, these compounds contribute antioxidant and antiinflammatory benefits upon consumption. Their use in meat products, edible oils, and bakery foods demonstrates how nutraceutical preservatives can replace synthetic additives while enhancing nutritional quality (Branen et al., 2002). Furthermore, many plant-derived and fermentation-based preservatives exhibit broad-spectrum antimicrobial activity against spoilage and pathogenic microorganisms, thereby improving food safety during storage. Their increasing application in functional and minimally processed foods reflects the shift toward natural preservation strategies that simultaneously support product stability and consumer health.

Table 4.1 Classification of nutraceutical-based food additives with technological and health functions.

Nutraceutical class	Examples	Technological role	Health relevance	Typical food applications
Preservatives	Rosemary extract, catechins	Oxidation and microbial control	Antioxidant, antiinflammatory	Meat, bakery, oils
Colorants	Curcumin, anthocyanins	Color stabilization	Antioxidant	Beverages, dairy
Sweeteners	Steviol glycosides	Sweetness	Glycemic control	Diabetic foods
Emulsifiers	Lecithin, fibers	Emulsion stability	Gut health	Dairy, beverages
Fortifiers	Omega-3, phytosterols	Nutritional enrichment	Cardiovascular health	Functional foods

4.4.2 Nutraceutical colourants

Natural pigments such as curcumin, anthocyanins, carotenoids, chlorophylls, and betalains are increasingly employed as colorants with added nutraceutical value. Beyond imparting color, these pigments exhibit antioxidant and antiinflammatory properties (Scotter, 2005). However, their sensitivity to light, heat, oxygen, and pH presents formulation challenges. Encapsulation technologies have improved pigment stability, enabling their application in beverages, dairy products, and functional foods without compromising color intensity or bioactivity (Shahidi & Ambigaipalan, 2015). In addition to their functional and esthetic roles, these bioactive pigments contribute to the overall health-promoting profile of foods by supporting antioxidant defense systems and protecting cellular components from oxidative stress. Consequently, the integration of nutraceutical colorants into food formulations represents an important approach in developing visually appealing products that also deliver added physiological benefits.

4.4.3 Nutraceutical sweeteners

Natural high-intensity sweeteners such as steviol glycosides and monk fruit extract are widely used to reduce sugar content while maintaining sweetness. These compounds provide negligible calories and do not significantly affect blood glucose levels, making them suitable for diabetic and weight-management foods (Magnuson et al., 2016). Sugar alcohols such as xylitol and erythritol also function as nutraceutical sweeteners, offering reduced caloric value and additional benefits such as dental caries prevention. Their multifunctional role illustrates how nutraceutical additives improve both sensory quality and metabolic outcomes. In addition to their role in reducing caloric intake, several nutraceutical sweeteners exhibit functional benefits such as prebiotic effects and improved glycemic control. Their incorporation into functional foods and beverages reflects the growing emphasis on developing products that balance palatability with metabolic health and dietary management. Moreover, the use of nutraceutical sweeteners supports the development of low-glycemic and reduced-sugar formulations that align with current public health recommendations for limiting excessive sugar intake. Continued research on their sensory optimization, metabolic responses, and long-term safety is expected to further expand their applications in next-generation functional foods and nutraceutical formulations.

4.4.4 Nutraceutical emulsifiers and stabilizers

Certain nutraceutical-derived compounds, including lecithin, dietary fibers, and protein hydrolysates, function as emulsifiers and stabilizers in food systems. These ingredients improve emulsion stability, viscosity, and mouthfeel while contributing to digestive health and cholesterol management (Imeson, 2009). For example, replacing synthetic emulsifiers in dairy beverages with fiber-based stabilizers improves product texture and simultaneously enhances dietary fiber intake, demonstrating the functional advantage of nutraceutical additives. In addition to improving physicochemical stability, these biofunctional ingredients contribute to the nutritional profile of foods by supporting gut health and modulating lipid metabolism. Their application in functional formulations reflects a growing preference for multifunctional ingredients that simultaneously enhance product stability, texture, and health-promoting value. Furthermore, the incorporation of nutraceutical-based emulsifiers and stabilizers supports the development of clean-label food products by reducing reliance on synthetic additives. Ongoing research into plant- and marine-derived biopolymers is expanding the range of natural ingredients capable of providing both technological functionality and added nutritional benefits in complex food systems.

4.4.5 Nutraceutical fortifying additives

Nutraceutical fortifying additives include omega-3 fatty acids, phytosterols, vitamins, and mineral–bioactive complexes incorporated into foods to enhance nutritional density. These additives support specific health claims related to cardiovascular health, bone health, and metabolic regulation (Magnuson et al., 2016). Ensuring stability and bioavailability during processing remains a critical challenge. Encapsulation and controlled-release systems are increasingly employed to protect sensitive nutraceuticals and maintain their functional efficacy. Furthermore, the strategic incorporation of nutraceutical fortifying additives enables the development of foods that address specific nutritional deficiencies and support preventive healthcare. These fortificants play a significant role in functional food design by delivering essential bioactive compounds without substantially altering sensory attributes or consumer acceptance. Advances in microencapsulation, nanoemulsion systems, and carrier-based delivery technologies have improved the protection, stability, and targeted release of sensitive nutraceuticals during processing and digestion. Such technologies also enhance the bioaccessibility and absorption of bioactive compounds within the gastrointestinal tract. Consequently, nutraceutical fortification has emerged as an important approach for developing next-generation foods that combine nutritional enrichment with measurable physiological benefits. The growing interest in personalized nutrition and health-oriented food products further reinforces the relevance of nutraceutical fortifying additives in modern food systems.

4.5 Safety, stability, and regulatory considerations

Despite their natural origin, nutraceutical additives require rigorous safety evaluation. Factors such as dosage, cumulative exposure, bioavailability, and interactions with food matrices must be assessed to ensure long-term safety (WHO & FAO, 1987). Regulatory authorities such as Codex Alimentarius, EFSA, and FSSAI require scientific evidence demonstrating both safety and technological justification. Health claims associated with nutraceutical additives must be substantiated through validated scientific data and comply with regulatory frameworks (Codex Alimentarius Commission, 2023). Furthermore, comprehensive toxicological assessments, including acute and chronic exposure studies, are essential to establish acceptable daily intake (ADI) levels and ensure consumer safety. Stability during processing, storage, and distribution is another critical factor, as degradation of bioactive compounds may reduce efficacy or lead to the formation of undesirable by-products.

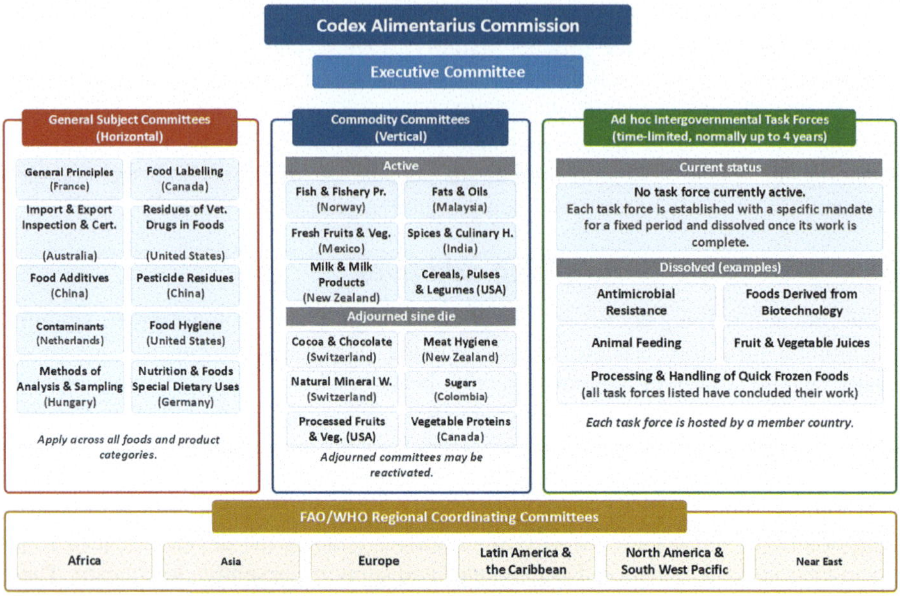

Note: Six regional coordinating committees support Codex work in their region. Committee statuses and host countries reflect the current Codex Procedural Manual; commodity committees adjourned sine die may be reactivated when new work arises.

Source: Compiled by the author based on FAO/WHO Codex Alimentarius Commission, Procedural Manual (latest edition), Rome: FAO/WHO; and the committee listings on the Codex Alimentarius website (www.fao.org/fao-who-codexalimentarius).

Figure 4.1 Organizational structure of the Codex Alimentarius Commission. Organizational structure of the Codex Alimentarius Commission illustrating the Executive Committee, horizontal committees addressing general food standards, commodity-specific committees, ad-hoc intergovernmental task forces, and FAO/WHO regional coordinating committees responsible for global food safety standardization.

Standardization of raw materials and quality control measures are therefore necessary to maintain consistency in nutraceutical additive formulations. Advances in analytical techniques and risk assessment models have improved the ability to monitor purity, potency, and potential contaminants in complex food systems. Transparent labeling and evidence-based health claims also play an important role in maintaining consumer trust and regulatory compliance. As the functional food sector continues to expand, harmonization of international regulatory frameworks will be crucial for facilitating global acceptance and responsible use of nutraceutical additives. The overall decision pathway involved in selecting nutraceutical food additives is illustrated in Fig. 4.2.

4.6 Technological and commercial challenges

The application of nutraceutical food additives faces several challenges, including variability in raw material composition, higher production costs, sensory interactions, and limited scalability of extraction processes. Regulatory complexity related to health claims further complicates commercialization (Table 4.2) (Baines and Seal, 2012).

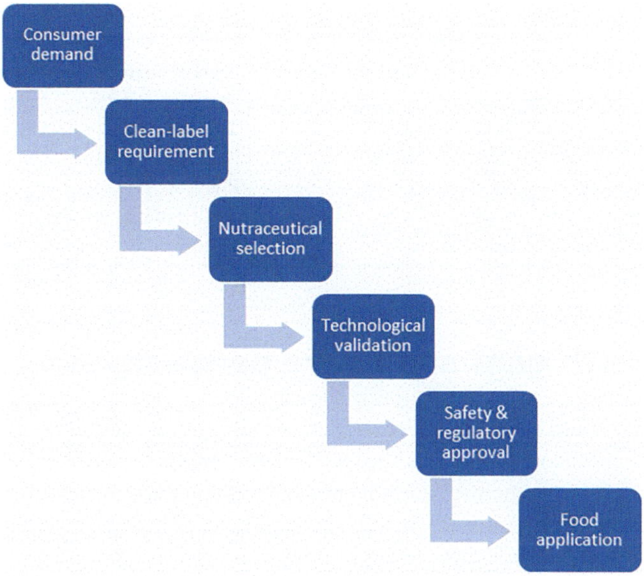

Figure 4.2 Decision pathway for selection and application of nutraceutical food additives. The figure illustrates the sequential decision-making process involved in selecting nutraceuticals as food additives, beginning with consumer demand and clean-label requirements, followed by nutraceutical selection, technological validation, safety and regulatory approval, and final application in food products. Despite their functional and health advantages, the widespread adoption of nutraceutical food additives is influenced by several technological and commercial constraints.

In addition, maintaining consistency in the quality and concentration of bioactive compounds remains a significant challenge due to variations in plant sources, seasonal factors, and processing conditions. The integration of nutraceutical additives into complex food matrices may also influence flavor, color, and texture, requiring careful formulation strategies to ensure consumer acceptance. Technological limitations in large-scale extraction, purification, and stabilization of bioactive compounds can further increase production costs and affect industrial feasibility. Moreover, demonstrating scientifically validated health benefits demands extensive research, clinical evidence, and compliance with stringent regulatory requirements. These factors collectively influence market entry timelines and commercial competitiveness. Addressing these challenges through advances in processing technologies, supply chain standardization, and collaborative research will be essential for the successful commercialization of nutraceutical food additives.

Addressing these challenges requires advances in standardization, formulation technology, and interdisciplinary collaboration between food technologists, nutrition scientists, and regulatory experts. A comparison between conventional food additives and nutraceutical food additives is presented in Table 4.2, highlighting their functional and consumer-oriented differences.

Table 4.2 Comparison between conventional food additives and nutraceutical food additives.

Aspect	Conventional additives	Nutraceutical additives
Primary function	Technological	Technological + health
Source	Often synthetic	Natural/bioactive
Consumer perception	Neutral/negative	Positive
Health contribution	None	Antioxidant, metabolic, gut health
Clean-label suitability	Limited	High

4.7 Research gaps and future perspectives

Future research should focus on improving the stability, bioavailability, and synergistic functionality of nutraceutical food additives. Greater emphasis is needed on food-based validation of health benefits, dose–response relationships, and integration with nonthermal processing and advanced packaging technologies (Rahman, 2007). As food systems evolve toward preventive nutrition and sustainability, nutraceutical-based food additives are expected to play a central role in bridging food technology with public health objectives. The integration of nutraceuticals as food additives therefore represents a critical strategy for developing safe (Akoh & Min, 2017; Codex Alimentarius Commission, 2023; Food Safety and Standards Authority of India FSSAI, 2011; Shahidi, 2012; World Health Organization WHO, Food and Agriculture Organization FAO, 1987; European Food Safety Authority EFSA, 2021), stable, and health-oriented food products within modern food systems.

AI disclosure

During the preparation of this work, the author(s) used Open AI in order to check grammar and spelling. After using this tool/service, the author(s) reviewed and edited the content as needed and take(s) full responsibility for the content of the publication.

References

Akoh, C.C., Min, D.B., 2017. Food Lipids: Chemistry, Nutrition, and Biotechnology, fourth ed. CRC Press, Boca Raton, FL. Available at https://doi.org/10.1201/9781315121069.

Baines, D., Seal, R., 2012. Natural Food Additives, Ingredients and Flavourings. Woodhead Publishing, Cambridge, UK. Available at https://doi.org/10.1533/9780857095725.

Branen, A.L., Davidson, P.M., Salminen, S., Thorngate, J.H., 2002. Food Additives, second ed. CRC Press, Boca Raton, FL. Available at https://doi.org/10.1201/9780203904030.

Codex Alimentarius Commission, 2023. General Standard for Food Additives (GSFA). FAO/WHO, Rome. Available at https://www.fao.org/fao-who-codexalimentarius/codex-texts/list-standards/en/.

Damodaran, S., Parkin, K.L., Fennema, O.R., 2007. Fennema's Food Chemistry, fourth ed. CRC Press, Boca Raton, FL. Available at https://doi.org/10.1201/9781420027976.

European Food Safety Authority (EFSA), 2021. Re-evaluation of food additives: principles and procedures. EFSA Journal 19(5), e06587. Available at https://www.efsa.europa.eu/en/topics/topic/food-additives.
Food Safety and Standards Authority of India (FSSAI), 2011. Food Safety and Standards (Food Products Standards and Food Additives) Regulations, as amended. FSSAI, New Delhi. Available at https://www.fssai.gov.in.
Imeson, A., 2009. Food Stabilisers, Thickeners and Gelling Agents Food Stabilisers, Thickeners and Gelling Agents. Wiley-Blackwell, United Kingdom, pp. 1–352. Available at http://onlinelibrary.wiley.com/book/10.1002/9781444314724, https://doi.org/10.1002/9781444314724.
Magnuson, B.A., Carakostas, M.C., Moore, N.H., Poulos, S.P., Renwick, A.G., 2016. Biological fate of low-calorie sweeteners. Nutrition Reviews 74 (11), 670–689. Available at https://doi.org/10.1093/nutrit/nuw032.
Rahman, M.S. (Ed.), 2007. Handbook of food preservation, second ed. CRC Press, Boca Raton, FL. Available at https://doi.org/10.1201/9781420017373.
Scotter, M.J., 2005. The chemistry and analysis of food colourants. Food Additives and Contaminants 22 (6), 557–565. Available at https://doi.org/10.1080/02652030500111905.
Shahidi, F., 2012. Nutraceuticals, functional foods and dietary supplements in health and disease. Journal of Food and Drug Analysis 20 (1). Available at https://doi.org/10.38212/2224-6614.2144.
World Health Organization (WHO), Food and Agriculture Organization (FAO), 1987. Principles for the Safety Assessment of Food Additives and Contaminants in Food. Environmental Health Criteria 70. WHO, Geneva. Available at https://apps.who.int/iris/handle/10665/37287.

Natural and nutraceutical preservatives

5

Sudipta Kumar Mohanty[1] *and Nagamani J.E.*[2]
[1]Department of Biotechnology, Padmashree Institute of Management and Sciences, Bengaluru, Karnataka, India, [2]Department of Food Science and Nutrition, Padmashree Institute of Management and Sciences, Bengaluru, Karnataka, India

5.1 Introduction

Food spoilage is caused mainly due to microbial growth, enzymatic activity and other chemical reactions such as oxidation. The pathogenic microorganisms and their toxins can lead to foodborne illnesses, affecting public health. According to Pawar et al. (2011), bacteria, yeasts, and molds utilize nutrients from food in the form of carbohydrates, proteins, and lipids, which leads to undesirable changes in food texture, flavor, color, and nutritional value. Enzymatic action further accelerates food spoilage by promoting browning, softening, and faster nutrient degradation. The oxidative processes cause rancidity and loss of sensory quality, particularly in lipid-rich foods (Adamczak et al., 2020; Fallah Ziarani et al., 2023; Mantzourani et al., 2025; Rajadurai et al., 2022; Yilmaz Atay, 2020; Mesías et al., 2021; Shaltout, 2024). Food spoilage not only reduces shelf life but also poses serious food safety concerns. Apart from microbial spoilage, improper storage, inadequate preservation methods, and use of excessive synthetic additives further compound the health risks. Implementation of effective preservation strategies is crucial in minimizing post-harvest losses, ensuring microbiological safety, maintaining nutritional quality, and supporting food security (Teshome et al., 2022).

The food preservatives used are generally categorized as synthetic and natural preservatives. These preservatives are widely used in food and beverages to prevent bacterial and fungal growth, increase food stability and nutritional quality, and prevent oxidation (Malek and Moslemi, 2025). Due to its huge demand, use of unauthorized food preservatives has increased dramatically and raises concern among consumers. Increasing awareness for a safe, nutritious, and minimally processed food has shifted interest of people toward natural preservatives (Shakil et al., 2022; Di Nunzio et al., 2022). Frequent use of Synthetic or chemical preservatives like butylated hydroxyanisole (BHA), sodium benzoate, and nitrites has raised serious health concern due to their potential toxicity and adverse impact. Researchers and the food industries are continuously looking for a safer natural alternative preferably from plants, animals, or microorganisms with health-promoting effects. Such preservatives will serve dual function and inhibit microbial food spoilage and oxidative degradation along with additional health benefits. Natural food preservatives have been found to exhibit diverse biological functions, including antioxidant,

Nutraceutical Applications in Food Processing and Preservation. DOI: https://doi.org/10.1016/B978-0-443-45784-5.00011-3

antimicrobial, and antiinflammatory activities. Plant-derived compounds such as polyphenols, flavonoids, essential oils, and tannins are well known for strong antimicrobial potential against foodborne pathogens. Similarly, bioactive compounds like lactic acid, citric acid, and acetic acid serve as natural preservatives by lowering the pH and inhibiting microbial growth in food. Carotenoids and tocopherols from plants act as potential antioxidants that prevent lipid oxidation and maintain the stability and enhance shelf life of fat-enriched foods. Apart from plant source, natural preservatives from microbes are of great interest. Compounds like nisin and bacteriocins produced by lactic acid bacteria provide protection of dairy and meat products from Gram-positive bacteria. Apart from food preservation, these compounds also profoundly add to the nutritional value of foods. Their antioxidant properties reduce oxidative stress in the human body and antiinflammatory, immune-boosting effects promote overall health.

5.2 Role of preservatives in food processing

The biocompatibility and safety make natural preservatives a promising alternative for synthetic food additives. Stability, sensory impact, and cost-effectiveness of natural preservatives play a critical role in large-scale food processing. Continued research and new technology for extraction and formulation have made the way for practical utility of these natural compounds in modern food preservation system. Increased awareness among people for the use of natural bioactive compounds will be a significant step toward ensuring food safety, extending shelf life, and promoting consumer health.

5.2.1 Antimicrobial action of food preservatives

The antimicrobial effect of natural food preservatives from plant sources is well documented. The hydroxyl (OH) group presents on phenolic compounds interacts with the microbial cell wall, causes structural disintegration, and increases cell permeability and leakage of cellular contents. Some phenolic compounds from plants interfere with the enzyme activity of normal metabolic process of microbes. Therefore, phenolic compounds disrupt biosynthesis process of nucleic acid within microbial cells. Gram-positive bacteria are reported to be more sensitive to phenolics than Gram-negative species due to the differences in cell wall composition. The phenolic compounds also induce oxidative stress by producing reactive oxygen species (Qu et al., 2019). Phenolic compounds can also exhibit antibiofilm activity on foodborne pathogen. Certain compounds also disrupt the growth of microbes by interfering the cell-to-cell communication, which is necessary for spread and growth of microorganisms.

5.2.2 Antioxidant principle of food preservatives

Natural preservatives have also proven to be powerful antioxidants, capable of scavenging free radicals. They prevent oxidative degradation of protein- and lipid-rich foods. The phenolic compounds donate hydrogen atoms or electrons to neutralize the reactive oxygen species (ROS) and prevent lipid peroxidation in food. Extracts from green tea, rosemary, and grape seed have been currently in use as strong lipid stabilizer in meat products, oils, and dairy-based products. By preventing rancidity, phenolics are also helpful in preservation of flavor, color, and nutritional value.

5.3 Preservation and food security

In the current era, food preservation is most crucial for food security as it ensures safety, availability, and usability for a longer duration. An irreplaceable amount of food wastage happens due to microbial contamination, enzymatic action, and improper storage conditions. Traditional food preservation methods such as drying, refrigeration, canning, and fermentation have been used to slow down spoilage to a greater extent. Many foods are produced seasonally in larger quantities and depend on the climatic conditions and geography. The preservation method allows surplus food to be stored and consumed during nonavailability or adverse climatic conditions like droughts and floods, ensuring a no shortage of food supply. Efficient preservation methods of foods promote longer shelf life and contribute to store for longer duration and transport to far places. During natural disasters or pandemics, preserved foods play a vital role in feeding affected populations. By practicing food preservation techniques, food producers also get benefited by reducing postharvest losses. This also contributes to stronger agricultural systems and better national food security. Thus, food preservation is considered a key pillar of food security.

5.4 Synthetic food preservatives and associated concerns

5.4.1 Common synthetic food preservatives and its importance

Synthetic food preservatives are added to foods to reduce spoilage caused by microorganisms, oxidation, and enzymatic action. They play a critical role by extending shelf life of foods, maintaining quality, and ensuring food availability for processed and packaged foods, a modern day's requirements. The Synthetic preservatives are broadly classified based on their nature of action like antimicrobial, antioxidant, or antienzymatic agents (Anand and Sati, 2013). Antimicrobial synthetic food preservatives such as sodium benzoate, potassium sorbate, sodium nitrite, sodium nitrate, and sulfites inhibit the growth of bacteria, yeasts, and molds on foods. These kinds of preservatives are widely used in fruit juices, jams, pickles, sauces, beverages, dried fruits, and processed meat products (Gharu, 2016; Sharma and Rajput, 2023).

Chemical preservatives like nitrites and nitrates are frequently used in meat-based products to prevent microbial growth and maintain color. Nitrates and nitrites are important preservatives in cured meat products as they prevent growth of harmful microbes like *Clostridium botulinum.* These preservatives are also known for maintaining color and flavor of meat and meat-based packaged products.

Antioxidant preservatives such as BHA, BHT, propyl gallate, and TBHQ are also commonly used to delay oxidative rancidity in fats and oils. These antioxidants act by interrupting free radicals and helps retaining the sensory and nutritional quality of bakery products, cereals, snacks, and edible oils (Anand and Sati, 2013; Rajadurai et al., 2022). Preservatives like citric acid and erythorbic acid block the enzyme activity and are used to reduce enzymatic browning in fruits and vegetables after harvest (Sharma and Rajput, 2023).

5.4.2 Safety and health concerns of synthetic preservatives

Despite having regulatory approval, many synthetic preservatives have been reported with potential health risks. Synthetic preservatives, nitrates, and nitrites can react with amines in the gastrointestinal tract to form N-nitroso compounds, which are known to be a potent carcinogen. Research shows there are positive associations between long-term consumption (Anand and Sati, 2013) of nitrite-treated foods and risks of colorectal, gastric, and esophageal cancers (Rajadurai et al., 2022). Sulfites are proven to cause hypersensitivity reactions, especially in asthmatic persons. Adverse effects like bronchial constriction, skin irritation, headaches, abdominal pain, and allergic response are also reported by several researchers (Aldabayan, 2025; Joute et al., 2016). Sulfite sensitivity can cause severe respiratory symptoms even at low exposure levels in susceptible individuals (Rajadurai et al., 2022; Ukwo et al., 2022). Sodium benzoate is reported to have a positive correlation to allergic reactions, potential genotoxic effects, and hyperactivity in children. Studies have reflected that sodium benzoate may cause chromosomal aberrations and DNA damage. For example, sodium benzoate when consumed in combination with ascorbic acid may lead to benzene formation, which has adverse health effects and causes cancer and hematological disorder (Gharu, 2016; Rajadurai et al., 2022). Animal studies indicate that excessive intake of synthetic antioxidants results in liver toxicity, endocrine disruption, and tumor formation (Anand and Sati, 2013).

5.4.3 Environmental impact

Another side of safety concern is the production, use, and disposal of synthetic preservatives, which may contribute to environmental hazards. Though synthetic food preservatives have significantly contributed to modern food preservation, a safer alternative always remains in focus. Public awareness, increasing health, environmental, and consumer concerns have encouraged stricter regulations by the government. Therefore, there is a growing shift toward natural preservatives and safer preservation technologies are observed in the recent past (Mesías et al., 2021).

5.4.4 Consumer perception and clean-label demand

Many people prefer "clean-label," natural or synthetic preservative-free foods due to fears about chemical exposure and health hazards. In this regard, regulatory authorities and government bodies have set acceptable daily intake (ADI) limits. However, improper usage, lack of knowledge and awareness, and repetitive intake from multiple food sources can pose safety threats. Consumer resistance to chemical preservatives has encouraged manufacturers to explore more natural sources of preservatives like plant extracts, essential oils, organic acids, and bio-preservatives. Clean-label products are assumed to be safer, healthier, and more environmentally friendly among people (Sharma and Rajput, 2023).

5.5 Regulatory restrictions and usage limits

The use of synthetic preservatives is regulated by food safety authorities such as the Food and Drug Administration (FDA), Food Safety and Standards Authority of India (FSSAI), European Food Safety Authority (EFSA), and largely by the World Health Organization (WHO). These organizations establish the use of preservatives as ADI values and maximum permissible limits to ensure consumer safety. For instance, benzoates are generally limited to approximately 200 ppm in acidic foods; nitrates and nitrites have more strict limits due to its harmful and carcinogenic potential (Sharma and Rajput, 2023).

The European Union assigned unique numbers to preservatives, E-numbers (E200–E299), and provided guidelines that only approved preservatives may be used in food under specified conditions. Several synthetic preservatives have been restricted or banned in some countries following toxicological evidence and to reduce the toxicity level (Anand and Sati, 2013). Continuous monitoring and reassessment of food quality are essential to balance food safety and public health (Rajadurai et al., 2022). Synthetic preservatives play an essential role in extending the shelf life and safety of modern food products. Growing consumer awareness, advanced analytical techniques, and knowledge about food quality have accelerated the clean-label movement, driving demand for natural alternatives. In this scenario, regulatory agencies continue to impose strict usage limits and revise safety standards to minimize risks and to meet consumer demand.

5.6 Natural and nutraceutical preservatives

5.6.1 Concept of natural and nutraceutical food preservatives

Food preservatives are the compounds added to food to avoid spoilage caused by microbes, oxidation reaction, and undesirable enzymatic and chemical reactions. The natural food preservatives are obtained from plants, animals, or microorganisms that possess an inherent antimicrobial or antioxidant properties. The concept of natural food preservation is rooted back to traditional practices such as drying, salting, fermentation, and the use of different herbs and spices (Bubai, 2023). Nutraceutical food preservatives are more advantageous with the concept of combining preservative action with health-promoting effects. Many natural preservatives contain bioactive compounds like phenolics, flavonoids, terpenoids, peptides, and organic acids. These compounds are actively involved in antioxidant, antimicrobial, antiinflammatory, and immune-boosting potential. Thus, nutraceutical preservatives serve a dual function, food preservation as well as promotion of human health (Thakre, 2023; Teshome et al., 2022).

5.6.1.1 Role of natural preservatives in food preservation and health promotion

Natural preservatives act primarily through three mechanisms:

Antimicrobial action—inhibits and destroys bacteria, yeasts, and molds.

Antioxidant action—prevents or delays lipid oxidation and rancidity.

Antienzymatic action—inhibits enzymes, responsible for browning, ripening, and degradation.

Plant-derived phenolic compounds and essential oils disrupt microbial membranes and metabolic pathways. Animal-derived peptides and enzymes damage microbial cell walls or limit nutrients for microbes. Microbial metabolites such as bacteriocins inhibit the growth of microorganisms (Pawar et al., 2011; Teshome et al., 2022). Natural preservatives also help to maintain aroma, taste, texture, and nutritional value, and are widely used in raw, cooked, fermented, and minimally processed food items.

Apart from food preservation, natural preservatives also contribute to health-promoting activity. Antioxidants are proven to reduce oxidative stress and cellular damage. Antimicrobial compounds reduce the growth and exposure of foods to foodborne pathogens. Certain preservatives of animal and microbial origin enhance the immune responses and gut health. Natural preservatives are associated with fewer toxicological risks as compared to synthetic preservatives. However, excessive intake or frequent use of some compounds may cause mild or severe adverse effects (Thakre, 2023; Khan et al., 2024).

5.7 Sources of natural preservatives

Natural food preservatives are derived mainly from three sources: plant sources including essential oils, animal sources, and microbial sources (Fig. 5.1).

5.7.1 Plant-based food preservatives

Plants are the most abundant source of natural preservatives. Herbs, spices, and essential oils contain secondary metabolites like phenolics, flavonoids, terpenoids, alkaloids, and organic acids with strong antimicrobial and antioxidant potential. These preservatives play a critical role in food preservation, which includes prolonged shelf life, improving or maintaining nutritional quality, and food safety. The demand for natural and safer alternatives to synthetic food preservatives has directed focus toward plant-derived compounds as a natural source. Among all, phenolic compounds have been most widely used as effective natural preservatives due to their potent antioxidant, antimicrobial, and enzyme inhibitory action. Phenolic compounds have an aromatic ring with one or more hydroxy groups attached to it and functional derivatives like esters. These compounds are commonly found across the

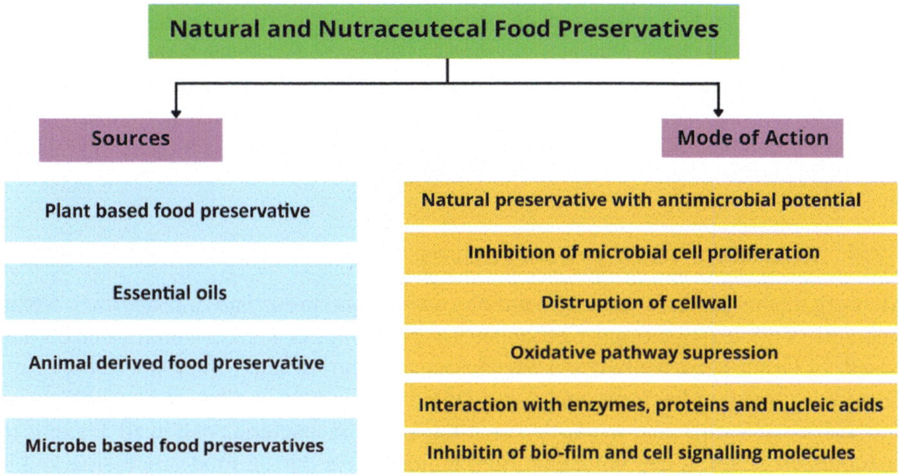

Figure 5.1 Classification of natural and nutraceutical preservatives. Different sources and modes of action of preservatives.

plant kingdom and greatly contribute to color, flavor, and defense mechanism in plants. The phenolic compounds are classified as phenolic acids (e.g., gallic acid, caffeic acid, and ferulic acids), flavonoids (quercetin, catechin, and naringenin), tannins, coumarins, and lignans. Natural sources of these compounds are fruits, vegetables, medicinal plants, herbs, and trees. Bioactive compounds from different medicinal plants act as natural preservatives (Pawar et al., 2011). Few examples like curcuminoids in *Curcuma longa* (turmeric) show antibacterial and antifungal activity (Adamczak et al., 2020). Cinnamaldehyde in *Cinnamomum zeylanicum* (cinnamon) inhibits bacteria and fungi growth. Thymol in *Ocimum* species (basil) exhibits antimicrobial effects (Mantzourani et al., 2025). Capsaicin in *Capsicum* species inhibits spoilage microorganisms. These compounds act mainly by disrupting the microbial cell wall and inhibiting the enzyme activity of regular metabolic processes (Julizan et al., 2023).

5.7.2 Essential oils from plants as food preservatives

Essential oils from plants like clove, thyme, oregano, eucalyptus, and coriander possess strong antimicrobial and antioxidant activities. They are mainly used against foodborne pathogens and to reduce oxidative deterioration. The strong aroma and volatility may limit their application, but they are more appealing due to their natural origin and health-promoting activity (Teshome et al., 2022; Noshirvani and Keng, 2024).

5.7.3 Animal-derived food preservatives

Certain animal-derived products act as natural preservatives due to their biological functions: Chitosan, derived from crustacean shells, inhibit growth of bacteria, fungi, and algae. Lysozyme, found in egg white and milk, inhibits growth of mainly Gram-positive bacteria in food. Lactoferrin, a protein present in milk and secretions, has antibacterial, antiviral, antifungal, and immune modulating potential. Defensins, an antimicrobial peptide, are reported with broad-spectrum antimicrobial activity. Animal-derived preservatives found their place in food, pharmaceutical, and cosmetic formulations due to their biocompatibility, effectiveness, and health safety concerns (Teshome et al., 2022).

5.7.4 *Microbial food preservatives*

Microorganisms produce several metabolites with food preservative properties. Nisin, a natural peptide (bacteriocin) produced by *Lactococcus lactis*, which controls the growth of Gram-negative bacteria *Staphylococcus aureus* and *Streptococcus* species, other spore-forming bacteria, and fungi (Field et al., 2016; Fallah et al., 2023), is a heat-stable molecule that works by creating pores in bacterial cell wall leading to bacterial cell death. It can withstand high temperatures during food processing and is used in various dairy-based products like cheese, other dairy products, and meat (Abd-Elhamed et al., 2024).

Natamycin, a natural antifungal agent produced by *Streptomyces natalensis*, is used to prevent the growth of yeast and molds. It is used as a safe preservative in cheese, yogurt, and baking items to extend shelf life without altering color, texture, and flavor of food.

Reuterin (3-hydroxypropionaldehyde or 3-HPA), a low molecular weight nonprotein, is a broad-spectrum antimicrobial compound produced by *Limosilactobacillus reuteri*, a lactic acid bacteria. It is used in different dairy-based products, which control *Listeria*, *Salmonella*, and *E. coli* in milk and cheese. It is considered a powerful antimicrobial compound offering a safer and bio-based alternative to synthetic preservatives.

Organic acids like lactic acid, acetic acid, propionic acid, and citric acid are natural food preservatives produced by lactic acid bacteria also known as generally recognized as safe (GRAS). These organic acids inhibit the growth of food-spoiling microbes by releasing protons ($H+$) and altering intracellular pH, promote ATP synthesis, protein synthesis, and DNA replication, and ultimately leading to cell death.

Microbial preservatives are also used in fermented foods, where they enhance shelf life, nutraceutical and sensory quality (Pawar et al., 2011; Thakre, 2023). Natural and nutraceutical food preservatives play a vital role in modern food systems by ensuring safety, extending shelf life, and promoting health in different ways. In addition to food preservation and shelf life, their nutraceutical properties provide health benefits, making them more popular among people (Pawar et al., 2011; Teshome et al., 2022; Khan et al., 2024)

5.8 Classification of natural preservatives based on their function

Based on the functional role, natural preservatives are broadly classified into antimicrobial preservatives and antioxidant preservatives. This functional classification is widely accepted in food preservation literature (Pawar et al., 2011; Bubai, 2023; Teshome et al., 2022).

5.8.1 Natural preservatives with antimicrobial potential

The antimicrobial action of natural preservatives is due to disruption of microbial cell wall, inhibition of enzyme activity, nucleic acid synthesis, and alteration of intracellular pH (Pawar et al., 2011; Teshome et al., 2022). The antimicrobial preservatives, including bioactive compounds like cinnamon (cinnamaldehyde), clove (eugenol), thyme (thymol), oregano (carvacrol), turmeric (curcumin), garlic (allicin), and basil, have been reported to exhibit strong antimicrobial potential against foodborne pathogens (Pawar et al., 2011; Teshome et al., 2022). These compounds are commonly used to control microbial contamination in food. Animal-derived antimicrobial preservatives like lysozyme, lactoferrin, defensins, and chitosan are used in fish, poultry, meat, and some vegetables preservation (Karmakar et al., 2022). Lysozyme from egg white and milk acts against Gram-positive bacteria by hydrolyzing peptidoglycan in the bacterial cell wall (Khorshidian et al., 2022). Similarly chitosan obtained from crustacean shells exhibits broad-spectrum antimicrobial activity against bacteria, fungi, and yeasts and is used in preservation and preventing oxidation of fruits, vegetables, meat, sea foods, and dairy products (Pawar et al., 2011; Yilmaz, 2020). Microbial antimicrobial preservatives like bacteriocins, organic acids, and bioactive peptides produced during fermentation are reported to be very efficient and economical food preservatives. Nisin, reuterin, and natamycin are a few well-known antimicrobial preservatives that inhibit spoilage of foods (Pawar et al., 2011; Teshome et al., 2022).

5.8.2 Antioxidant natural preservatives

Antioxidants from natural sources act as food preservatives by delaying food deterioration caused by oxidation. The oxidation of lipids in food results in rancidity, alters flavors, discoloration, and affects nutritional quality in food. These preservatives quench free radicals, chelate metal ions, and interrupt oxidative chain reactions (Bubai, 2023; Sharma and Rajput, 2023). The natural antioxidants from plants are the most abundant and include phenolic compounds, flavonoids, tannins, carotenoids, and vitamins such as ascorbic acid (vitamin C) and tocopherols (vitamin E). Extracts from different plants like rosemary, green tea, clove, grape seed, and citrus fruits have been reported by many researchers to exhibit strong antioxidant activity (Bubai, 2023; Teshome et al., 2022). Animal-derived antioxidants such as lactoferrin and certain bioactive peptides also contribute to antioxidant activity in food by binding to metal ions (Pawar et al., 2011). Microbial metabolites produced during fermentation may also indirectly enhance antioxidant activity by stabilizing food matrices and reduce oxidative damage during food storage (Teshome et al., 2022). These natural preservatives also exhibit dual functionality of both antimicrobial and antioxidant properties. These preservatives are highly valuable for clean-label, functional, and nutraceutical food applications. Their use offers safer and health-promoting alternatives to synthetic preservatives (Pawar et al., 2011; Sharma and Rajput, 2023; Bubai, 2023; Teshome et al., 2022).

5.9 Mechanism of action of food preservatives

Food preservatives play a critical role in increasing shelf life, ensuring microbial safety, and maintaining sensory and nutritional quality of food. Most of the time, food spoilage is caused due to microbial growth, oxidative degradation of lipids, enzymatic reactions, and physicochemical instability. Food preservatives act through diverse biochemical and cellular mechanisms to overcome the food spoilage and inhibit microbial growth (Pawar et al., 2011; Teshome et al., 2022). Preservatives of natural or synthetic origin function rarely through a single pathway. They exert multiple ways like inhibition of microbial cell growth, membrane disruption, interference with enzyme activity, proteins and nucleic acids synthesis, inhibition of quorum sensing pathway and biofilm formation, loss of cell wall integrity, and intervenes with oxidative pathways (Malek and Moslem, 2025; Lemoni et al., 2023). These mechanisms must be explored for optimizing preservative efficacy and minimizing health risks.

5.9.1 Microbial cell inhibition

Inhibition of cell proliferation and cellular damage is the fundamental mechanism by which preservatives prevent food spoilage. Many preservatives prevent microbial growth by creating an altered intracellular environment that limits metabolism and cell division. Weak organic acids such as benzoic acid, sorbic acid, lactic acid, and acetic acid act in their undissociated form, which allows them to diffuse across microbial cell membranes (Pawar et al., 2011). Once it gets inside the cell, these acids dissociate and release protons, which in turn lower intracellular pH. This results in disruption of enzymatic activity, interferes with proton motive force, inhibits nutrient transport, and reduces ATP synthesis, which leads to bactericidal effects (Thakre, 2023). The effectiveness of this mechanism is strongly dependent on food pH, making an acidic environment particularly effective in fruit juices, fermented products, and pickles (Bubai, 2023).

5.9.2 Disruption of the cell membrane

Disruption of the microbial cell wall is another important mechanism associated with many natural preservatives, like plant-derived compounds. Essential oils and phenolic compounds like thymol, carvacrol, eugenol, cinnamaldehyde, and flavonoids get integrated into the phospholipid bilayer of microbial membranes. This integration induces membrane permeability, causes leakage of ions and cytoplasmic constituents, and leads to cell death (Teshome et al., 2022; Lemoni et al., 2025). Chitosan disrupts membranes due to electrostatic interactions between its positively charged amino groups and negatively charged microbial surfaces, which alter membrane integrity and permeability (Teshome et al., 2022). Synthetic preservatives, sulfites, also damage membranes indirectly by disturbing redox balance and membrane-associated proteins (Thakre, 2023). Gram-negative bacteria often exhibit greater resistance to food preservatives due to the presence of an outer membrane, whereas Gram-positive bacteria generally show more susceptibility (Lemoni et al., 2025).

5.9.3 Interaction with enzymes, microbial proteins, and nucleic acids

Many food preservatives exert antimicrobial effects by interacting with key enzymes, proteins, and nucleic acids and interfere with essential cellular processes. Sulfur dioxide and sulfites react with sulfhydryl (–SH) groups of microbial enzymes, which leads to enzyme inactivation and disruption of carbohydrate metabolism in yeasts and molds (Thakre, 2023). Phenolic compounds from plants can denature microbial proteins by disrupting hydrogen bonds and changing protein conformation. Some preservatives interfere with nucleotide (DNA and RNA) synthesis, which inhibits cell replication and transcription process (Pawar et al., 2011; Lemoni et al., 2023). Bacteriocins, nisin, exhibit a dual mode of action; they bind to lipid and inhibit peptidoglycan synthesis. At the same time, it also creates pores in the cytoplasmic membrane. This results in loss of cellular homeostasis and leads to rapid microbial cell death (Pawar et al., 2011).

5.9.4 Quorum sensing and biofilm Inhibition

Several studies highlighted quorum sensing (QS) mode of inhibition as an important mechanism of natural food preservatives. Quorum sensing is a cell-to-cell communication system that regulates the microbial pathogenicity, enzyme secretion, and biofilm formation. Biofilms enhance microbial resistance to preservatives and environmental stresses (Lemoni et al., 2025). Plant-derived bioactive compounds such as flavonoids, phenolic acids, terpenes, and tannins inhibit quorum sensing by interfering with signaling molecules (autoinducers) and block receptor binding and QS-related gene expression. The lack of biofilm formation capability weakens microbial defense mechanisms and increases susceptibility to antimicrobial agents like preservatives (Lemoni et al., 2025; Teshome et al., 2022). By disrupting quorum sensing, natural preservatives reduce microbial virulence without killing cells directly. Reduction in virulence also lowers the risk of resistance development and enhances the efficiency of preservatives in food.

5.9.5 Loss of cell wall integrity

The microbial cell wall is a critical target for several preservatives. Lysozyme hydrolyzes β-1,4-glycosidic linkages in peptidoglycan, weakening the cell wall and causing osmotic lysis, particularly in Gram-positive bacteria (Pawar et al., 2011). Bacteriocins and antimicrobial peptides inhibit cell wall biosynthesis by targeting peptidoglycan precursors, resulting in structural instability. Loss of cell wall integrity leads to increased susceptibility to osmotic stress, leakage of cytoplasmic contents, and eventual cell death (Teshome et al., 2022).

5.9.6 *Oxidative pathway suppression*

Oxidative deterioration is a major cause of food spoilage in foods with high lipid content. The antioxidant preservatives suppress oxidative pathways by neutralizing free radicals. Another mode of action is chelating metal ions and interrupting lipid oxidation chain reactions (Bubai, 2023). Phenolic compounds from plants, flavonoids, tocopherols, and carotenoids act as primary antioxidants, which donate hydrogen atoms or electrons and neutralize free radicals. Ascorbic acid behaves as a secondary antioxidant by regenerating oxidized antioxidants and scavenges oxygen (Thakre, 2023; Teshome et al., 2022). The enzymatic browning can be reduced by sulfur dioxide, which inhibits polyphenol oxidase activity, reduces quinones to phenolic compounds, and, ultimately, preserves color and sensory quality of food (Thakre, 2023).

5.9.7 *Synergistic effects in food matrices*

Preservatives also exhibit synergistic effects when combined with other preservatives, processing conditions, and food components. Organic acids enhance the activity of bacteriocins by increasing membrane permeability and facilitating bacteriocin entry into microbial cells (Pawar et al., 2011). Antioxidants like ascorbic acid, in combination with phenolics, improve the oxidative stability of food (Bubai, 2023). Hurdle technology is integrated with different preservation parameters like pH reduction, refrigeration, and packaging to get rid of microbial contamination even at lower concentrations of preservatives (Thakre, 2023). Sometimes, it is observed that complex food matrices of lipids and proteins may bind to preservatives and reduce their bioavailability (Teshome et al., 2022).

5.10 Nutraceutical benefits of natural preservatives

Natural food additives, derived from plants, animals, and microbes, are gaining significant attention due to their dual function as food preservatives and health-promoting activity. Apart from shelf-life enhancement and microbiological safety, these additives also show tremendous potential like antioxidant, antiinflammatory, immunomodulatory, and gut-protective properties. Unlike synthetic preservatives, natural food preservatives are often considered safer due to their traditional dietary use and lower toxicological concern. Many of these preservatives are also recognized as functional ingredients, which align with consumer demand for clean-label and health-oriented food products (Dávalos et al., 2022).

5.10.1 *Antioxidant activity*

Natural food additives often exhibit significant antioxidant potential due to the presence of polyphenols, flavonoids, phenolic acids, carotenoids, tocopherols, and vitamins. These preservatives play a crucial role in preventing oxidative deterioration of foods as well as oxidative stress in humans. Pawar et al. (2011) reported that plant-based preservatives, including herbs, spices, and essential oils, act as effective antioxidants by scavenging free radicals, chelating metal ions, and interrupting lipid peroxidation. The natural antioxidants derived from rosemary, clove, oregano, turmeric, green tea, grape seed, and fruit by-products are proven of delaying rancidity, stabilizing color, and prevent nutrient loss in food (Teshome et al., 2022). These antioxidants from natural sources also contribute to health promotion by reducing oxidative stress associated with chronic diseases like cardiovascular disorders, cancer, and neurodegenerative disorders. Several natural antioxidants are proven to be superior to synthetic antioxidants like BHT and BHA, offering better food safety and consumer acceptance.

5.10.2 Antiinflammatory effects

Antiinflammatory effects of natural preservatives are reported due to the presence of bioactive phytochemicals like polyphenols, flavonoids, terpenoids, and phenolic acids. Certain plant-derived additives, including turmeric, clove, cinnamon, ginger, oregano, and thyme, contain compounds like curcumin, eugenol, cinnamaldehyde, and carvacrol that have the capability of modulating inflammatory pathways. These compounds are involved in the synthesis of proinflammatory mediators by suppressing key enzymes such as cyclooxygenase (COX) and lipoxygenase (LOX) and downregulating nuclear factor-κB (NF-κB) signaling, which plays a central role in chronic inflammation reaction. The regular dietary exposure to such natural additives through preserved foods can help to reduce low-grade inflammation associated with metabolic disorders, gastrointestinal diseases, and oxidative related stress. Unlike synthetic additives, the natural preservatives provide antiinflammatory benefits and contribute to food safety and health promotion (Pawar et al., 2011; Teshome et al., 2022).

5.10.3 Gut health and immune modulation of natural food additives

Natural food preservatives play an important role in promoting gut health and modulating immune function through their selective antimicrobial, prebiotic, and bioactive properties. Natural additives such as organic acids, bacteriocins (e.g., nisin), chitosan, lysozyme, and plant-derived polyphenols inhibit pathogenic microorganisms and allow beneficial microbiota to survive in the gut (Pawar et al., 2011). This selective action helps to maintain the microbial balance of gastrointestinal flora, which is essential for digestion, nutrient absorption, and overall intestinal health. Research concludes that polyphenols and plant bioactives act as prebiotic-like compounds, promoting the growth of beneficial bacteria such as *Lactobacillus* and *Bifidobacterium*. Similarly, animal-derived food preservatives like lactoferrin enhance immune modulation by binding to iron, limiting pathogen growth, and stimulating host defense mechanisms. Through these combined effects, natural food additives contribute to reduced gut inflammation and strengthen mucosal immunity and overall enhancement of host immune response (Pawar et al., 2011; Teshome et al., 2022).

5.11 Challenges and limitations of natural food preservatives

Despite of widespread popularity and increasing interest in natural food preservatives due to their health benefits, several technological, economic, and regulatory challenges limit their widespread application in the food industry. The major limitations are related to stability, sensory impact, dose optimization, purity of preservative, efficacy, and cost (Pawar et al., 2011; Teshome et al., 2022).

5.11.1 Stability and sensory impact

One of the most significant challenges with natural food preservatives is their limited stability during food processing and storage conditions. Bioactive compounds such as polyphenols, essential oils, vitamins, and phenolic acids are often sensitive to heat, light, oxygen, and pH variations, leading to rapid degradation and reduced efficacy over time (Pawar et al., 2011). In addition to this, many natural preservatives like plant extracts and essential oils possess strong flavors, aromas, and colors, which can negatively affect the sensory quality of foods when used at effective concentrations. This sensory impact restricts their direct application and requires an encapsulation or masking strategy (Teshome et al., 2022).

5.11.2 Dose optimization and efficacy

Achieving the correct balance between preservation and acceptable sensory quality remains a challenge. Natural preservatives often require a higher concentration than synthetic additives to achieve antimicrobial or antioxidant effects. The complex food matrices, rich in lipids, fats, and proteins, bind to bioactive compounds and reduce their bioavailability (Teshome et al., 2022). Overdosage or higher concentration may result in undesirable taste, aroma, potential toxicity, and health concern while underdosage compromises with food safety. Moreover, variability of these preservatives in plant source, harvesting conditions, and extraction methods also leads to inconsistent composition and efficacy, making optimization a difficult task (Pawar et al., 2011).

5.11.3 Cost and scalability

The high cost of production is another bottle neck of natural food preservatives. Extraction, purification, and quality control of bioactive compounds from natural sources are often more expensive than the chemical preservatives. Seasonal availability of raw source materials, low extraction yields, and the need for advanced processing techniques further add to the costs (Selvam et al., 2025). These economic constraints limit the scalability of natural preservatives for large-scale production, low-cost food products, and pose big challenges for consistent industrial supply.

5.11.4 Regulatory and labeling challenges

Regulatory approval of natural food preservatives varies widely across countries and makes stringent compliance and commercial barriers. Although many natural additives are assumed to be safe, they do not get approval for food use and require extensive toxicological and safety evaluation parameters (Pawar et al., 2011). Differences in permitted levels, food categories, and labeling requirements complicate commerce at international level. Ambiguous terminologies such as "natural," "clean-label," and "functional" are related to regulatory interpretation and consumer perception, therefore, making natural food preservatives a significant challenge (Teshome et al., 2022). Natural food additives offer significant advantages in terms of health promotion and consumer acceptance, but challenges related to stability, efficacy, cost, and regulatory compliance limit their full-scale application. To address these limitations, technological innovations, such as encapsulation and hurdle technology, along with regulatory harmonization and cost-effective production strategies, are essential for the wide and sustainable use of natural food additives (Pawar et al., 2011; Teshome et al., 2022).

5.12 Conclusion and future prospects of natural food preservatives

The increasing demand for safe, sustainable, and health-promoting food products has made natural food preservatives promising alternatives to synthetic additives. Research suggests that the future of food preservation will increasingly rely on natural, multifunctional preservatives that provide food safety along with health benefits (Pawar et al., 2011; Teshome et al., 2022). One of the most significant future prospects lies in plant and biooriginated preservatives with enhanced efficacy. Advanced extraction techniques like supercritical fluid extraction, green solvents, enzyme-assisted extraction are expected to improve the yield, purity, and consistency of bioactive compounds from natural sources (Selvam et al., 2025). These technologies can overcome current limitations like variability and instability of natural preservatives. Encapsulation and nanodelivery systems represent another advancement for improving the stability, bioavailability, and controlled release of natural preservatives. Techniques such as nanoemulsions, liposomes, and biopolymer-based encapsulation can reduce undesirable sensory impact, protect bioactive compounds from quicker degradation, and enhance antimicrobial and antioxidant efficacy (Teshome et al., 2022). Such innovations will promote wider industrial adoption and enable the application of preservatives in diverse food matrices.

More research is also directed toward the use of hurdle technology. In hurdle technology, natural preservatives are combined with processing techniques like refrigeration, pH control, fermentation, and modified-atmosphere packaging.

This synergistic approach reduces higher dosage preservatives, minimizes sensory changes, and enhances overall food safety (Pawar et al., 2011). Their antioxidant, antiinflammatory, gut-modulating, and immune-enhancing properties contribute to disease prevention and overall well-being of humans (Teshome et al., 2022). Future research is likely to focus on finding the exact mechanism and linking specific preservative for a specific health care, which will strengthen the role of preservatives in preventive nutrition. Regulatory and consumer trends further support the future demand of natural food preservatives. Growing awareness of the potential health risks associated with synthetic additives and rising demand for clean-label and minimally processed foods are driving forces for higher demands (Pawar et al., 2011). Harmonization of regulatory frameworks and clearer guidelines for natural preservative approval and labeling will further facilitate their commercial use.

In conclusion, the future prospects of natural food preservatives are highly promising, which is driven by technological advancement, consumer demand, and health considerations. Continued research into stability enhancement, cost-effective production, synergistic applications, and regulatory standardization will be essential for their successful integration into modern food systems.

Natural food preservatives are substances derived from plant, animal, and microbial sources that extend the shelf life of foods while maintaining safety, quality, and nutritional value. Unlike synthetic preservatives, natural preservatives are more valued for their biological origin, functional properties, and consumer acceptance (Pawar et al., 2011; Teshome et al., 2022). Several researchers indicate natural preservatives act through multiple mechanisms rather than a single pathway. The mechanism of action by which natural and nutraceutical preservatives protects food are inhibition of microbial growth, disruption of microbial cell wall, modulating enzyme activity, interacting with proteins and nucleic acids, oxidative pathways, inhibition of quorum sensing and biofilm formation, and synergistic interactions within food matrices. These activities enhance the effectiveness of preservatives and reduce the risk of microbial resistance (Pawar et al., 2011; Teshome et al., 2022). Another important take-home message is the natural preservatives act as nutraceutical agents. Many natural preservatives exhibit antioxidant, antiinflammatory, gut-modulating, and immune-enhancing properties due to the presence of bioactive compounds such as polyphenols, flavonoids, terpenoids, peptides, and organic acids. Thus, natural preservatives not only protect foods from spoilage but also have health-promoting and disease prevention potential (Pawar et al., 2011). Replacement of synthetic additives with natural alternatives lowers consumer exposure to potentially harmful chemicals associated with hypersensitivity, toxicity, and long-term health risks. This aligns with consumer demand for safer, minimally processed foods and a healthy society (Teshome et al., 2022). The literature also highlights limitations of natural food preservatives like instability under processing conditions, strong sensory impact, variation in composition, higher cost, and regulatory challenges. These constraints hinder large-scale industrial adoption. The future progress depends on advances in advanced extraction process, encapsulation, nanodelivery, and hurdle technology to improve the quality and efficacy of preservatives (Pawar et al., 2011; Teshome et al., 2022). In conclusion, natural food preservatives represent a sustainable and health-oriented approach toward food preservation. The natural preservatives are most effective when viewed as multifunctional, synergistic, and system-dependent agents. The successful application requires integration of food science, nutrition, technology, and regulatory frameworks.

AI disclosure

During the preparation of this work, the authors used Open AI for the retrieval of literature summary and finding appropriate English words. After using this tool/service, the authors reviewed and edited the content as needed and take full responsibility for the content of the publication.

References

Abd-Elhamed, E.Y., El-Bassiony, T.A.E.R., Elsherif, W.M., Shaker, E.M., 2024. Enhancing Ras cheese safety: antifungal effects of nisin and its nanoparticles against *Aspergillus flavus*. BMC Veterinary Research 20 (1). Available at https://doi.org/10.1186/s12917-024-04323-1, https://bmcvetres.biomedcentral.com/.

Adamczak, A., Ożarowski, M., Karpiński, T.M., 2020. Curcumin, a natural antimicrobial agent with strain-specific activity. Pharmaceuticals 13 (7), 1–12. Available at https://doi.org/10.3390/ph13070153, https://www.mdpi.com/1424-8247/13/7/153/pdf.

Aldabayan, Y.S., 2025. Effect of artificial food additives on lung health—an overview. Medicina 61 (4), 684. Available at https://doi.org/10.3390/medicina61040684.

Anand, S.P., Sati, N., 2013. Artificial preservatives and their harmful effects: looking toward nature for safer alternatives. International Journal of Pharmaceutical Sciences and Research 4 (7), 2496–2501. Available at https://doi.org/10.13040/IJPSR.0975-8232.4(7).2496-01.

Bubai, B., 2023. Different types of food preservatives and its classification. Journal of Plant Pathology and Microbiology 14 (3), 1–8.

Dávalos, A., González-Castejón, M., Marina, M.L., Fraga, C.G., 2022. The role of natural antioxidants in food and health. Nutrients 14 (7), 1345.

Fallah Ziarani, M., Tohidfar, M., Mirjalili, M.H., 2023. Evaluation of antibacterial properties of nisin peptide expressed in carrots. Scientific Reports 13 (1). Available at https://doi.org/10.1038/s41598-023-49466-7, https://www.nature.com/srep/.

Field, D., O'Connor, R., Cotter, P.D., Ross, R.P., Hill, C., 2016. In vitro activities of nisin and nisin derivatives alone and in combination with antibiotics against Staphylococcus biofilms. Frontiers in Microbiology 7. Available at https://doi.org/10.3389/fmicb.2016.00508, http://journal.frontiersin.org/article/10.3389/fmicb.2016.00508/full.

Gharu, C.P., 2016. Synthetic food preservatives and their impact on human health. International Journal of Trend in Scientific Research and Development 1 (1), 161–165.

Joute, J.R., Chawhan, P., Rungsung, S., Kirthika, P., 2016. Food additives and their associated health risks. International Journal of Veterinary Science and Animal Husbandry 1, 1–5.

Julizan, N., Ishmayana, S., Zainuddin, A., Van Hung, P., Kurnia, D., 2023. Potential of *Syzygnium polyanthum* as natural food preservative: a review. Foods 12 (12), 2275. Available at https://doi.org/10.3390/foods12122275.

Karmakar, A., Ghosh, S., Islam, R., Rajak, P., 2022. Natural antimicrobials as alternative food preservatives in the dietary systems of human and livestock. Journal of Food Safety and Nutrition 12 (4), 119.

Khan, Z., Ahmad, M., Siddiqui, F., Rahman, S., 2024. Types of food preservatives, emerging technologies and impact on human health. International Journal of Creative Research Thoughts 12 (1), 1–10.

Khorshidian, N., Khanniri, E., Koushki, M.R., Sohrabvandi, S., Yousefi, M., 2022. An overview of antimicrobial activity of lysozyme and its functionality in cheese. Frontiers in Nutrition 9. Available at https://doi.org/10.3389/fnut.2022.833618, journal.frontiersin.org/journal/nutrition.

Lemoni, Z., Evangeliou, K., Lymperopoulou, T., Mamma, D., 2025. Incorporation of edible plant extracts as natural food preservatives: green extraction methods, antibacterial mechanisms and applications. Foods 14 (23), 4000. Available at https://doi.org/10.3390/foods14234000.

Malek, S., Moslemi, M., 2025. Application of natural preservatives in functional foods. Food Materials Research 5. Available at https://doi.org/10.48130/fmr-0025-0016, https://www.maxapress.com/data/article/fmr/preview/pdf/fmr-0025-0016.pdf.

Mantzourani, I., Daoutidou, M., Alexopoulos, A., 2025. The antimicrobial effect of thymol and carvacrol in combination with organic acids against foodborne pathogens in chicken and beef meat fillets. Microorganisms 13 (1), 182. Available at https://doi.org/10.3390/microorganisms13010182.

Mesías, F.J., Martín, A., Hernández, A., 2021. Consumers' growing appetite for natural foods: perceptions towards the use of natural preservatives in fresh fruit. Food Research International 150 (Pt A), 110749. Available at https://doi.org/10.1016/j.foodres.2021.110749.

Noshirvani, N., Keng, S.E., 2024. Essential oils as natural food preservatives: special emphasis on antimicrobial and antioxidant activities. Journal of Food Quality 2024 (1). Available at https://doi.org/10.1155/jfq/5807281.

Di Nunzio, M., Loffi, C., Montalbano, S., Chiarello, E., Dellafiora, L., Picone, G., et al., 2022. Cleaning the label of cured meat; effect of the replacement of nitrates/nitrites on nutrients bioaccessibility, peptides formation, and cellular toxicity of in vitro digested salami. International Journal of Molecular Sciences 23 (20), 12555. Available at https://doi.org/10.3390/ijms232012555.

Pawar, H.A., Shenoy, A.V., Narawade, P.D., Soni, P.Y., Shanbhag, P.P., Rajal, V.A., 2011. Preservatives from nature: a review. International Journal of Pharmacy and Phytopharmacological Research 1 (2), 78–88.

Qu, S., Yang, K., Chen, L., Liu, M., Geng, Q., He, X., et al., 2019. Cinnamaldehyde, a promising natural preservative against *Aspergillus flavus*. Frontiers in Microbiology 10. Available at https://doi.org/10.3389/fmicb.2019.02895, https://www.frontiersin.org/journals/microbiology#.

Rajadurai, M., Joseph, B., Soumya, S., 2022. Adverse effects of chemical preservatives: a review. Journal of Food and Nutrition 1 (1), 1–8. Available at https://doi.org/10.58489/2836-2276/002.

Selvam, T., Rahman, N.M.M.A., Olivito, F., Ilham, Z., Ahmad, R., Wan-Mohtar, W.A.A.Q.I., 2025. Agricultural waste-derived biopolymers for sustainable food packaging: challenges and future prospects. Polymers 17 (14). Available at https://doi.org/10.3390/polym17141897, http://www.mdpi.com/journal/polymers.

Shakil, M.H., Trisha, A.T., Rahman, M., Talukdar, S., Kobun, R., Huda, N., et al., 2022. Nitrites in cured meats: health risk issues and alternatives to nitrites. Foods 11 (21), 3355. Available at https://doi.org/10.3390/foods11213355.

Shaltout, F.A., 2024. Preservatives added to food preservation and their impact against the bacteria. Clinical Research and Clinical Reports 5 (1). Available at https://doi.org/10.31579/2835-8325/089.

Sharma, H., Rajput, R., 2023. The science of food preservation: a comprehensive review of synthetic preservatives. Journal of Current Research in Food Science 4 (2), 25–29.

Teshome, E., Mengistu, A., Debela, H., Tesfaye, T., 2022. Potentials of natural preservatives to enhance food safety and shelf life. Scientific World Journal 2022, 1–12.

Thakre, Y.G., 2023. Use of different types of preservatives in food industry with their benefits and adverse effects: a review. International Journal of Creative Research Thoughts 11 (12), h644–h648.

Yilmaz Atay, H., 2020. Antibacterial Activity of Chitosan-Based Systems Functional Chitosan: Drug Delivery and Biomedical Applications. Springer Singapore, Turkey, pp. 457–489. Available at http://doi.org/10.1007/978-981-15-0263-7, https://doi.org/10.1007/978–981-15–0263-7_15.

Natural sweeteners and nutraceutical benefits

6

B.K. Indu[1], *Shilpa P.*[2], *and Nagamani J.E.*[3]
[1]Department of Biotechnology, Padmashree Institute of Management and Sciences, Bengaluru, Karnataka, India, [2]Department of Nutrition and Dietetics, Padmashree Institute of Management and Sciences, Bengaluru, Karnataka, India, [3]Department of Food Science and Nutrition, Padmashree Institute of Management and Sciences, Bengaluru, Karnataka, India

6.1 Introduction

Excess dietary intake of free sugars, particularly from sucrose and sugar-sweetened foods, is strongly associated with increased risk of obesity, type 2 diabetes mellitus, dental caries, and cardiometabolic disorders. Accordingly, global public health bodies recommend limiting free sugar intake and encourage dietary shifts toward reduced sugar consumption (WHO, 2015). In response, the food industry has accelerated reformulation programs aimed at lowering added sugars while maintaining sensory acceptability and product stability.

Within this context, nutraceutical sweeteners have gained prominence. Nutraceutical sweeteners are defined here as sweetening agents that provide sweetness with either low/no glycemic impact and/or additional functional value (e.g., dental benefits, fermentation behavior, or bioactive potential). These sweeteners support "sugar-reduction" strategies and align with consumer demand for naturally sourced ingredients and clean-label products (Saraiva et al., 2020). However, sugar replacement is not merely a sensory substitution: sucrose contributes functional roles such as bulk, texture, browning, and preservation. Therefore, the scientific selection of sweeteners requires understanding of their technological performance and physiological effects (O'Donnell and Kearsley, 2012).

This chapter focuses specifically on (i) classification of sweeteners relevant to nutraceutical use, (ii) key formulation challenges during sugar replacement, (iii) major nutraceutical sweeteners used in modern products (Stevia, monk fruit, polyols, rare sugars), and (iv) nutraceutical/physiological benefits and safety considerations. Regulatory and market trends are presented only briefly to avoid overlap with other chapters in the book.

6.2 Classification of sweeteners relevant to nutraceutical formulation

Sweeteners can be classified based on their caloric contribution and metabolic fate. This framework is highly relevant to nutraceutical product development, particularly for foods designed for individuals with diabetes, obesity, and metabolic syndrome (Magnuson et al., 2016) (Fig. 6.1).

Nutraceutical Applications in Food Processing and Preservation. DOI: https://doi.org/10.1016/B978-0-443-45784-5.00014-9

COMPARISON OF SWEETENERS

Sweetener Types	Sugar	Polyols	High-Intensity Sweeteners	Rare Sugars
Caloric Content	4 kcal / gram	0 - 2.4 kcal / gram	0 kcal / gram	0.2 kcal / gram
Sweetness level	1x (Reference)	0.5x - 1x	100x - 300x	0.7x
Glycemic Impact	High	Low	No Impact	Minimal
Key Characteristics	High Calories, Rises Blood Sugar	Low Calories, Tooth Friendly	Very Sweet, No Calories	Low Calorie, Blood Sugar Friendly

Figure 6.1 Comparative classification of major sweetener categories used in nutraceutical food formulation. Comparison of major sweetener categories used in reduced-sugar and nutraceutical food formulations, including nutritive sugars, polyols (erythritol, xylitol), high-intensity sweeteners (stevia, monk fruit), and rare sugars. The figure highlights differences in caloric contribution, sweetness potency relative to sucrose, and glycemic impact, illustrating their functional roles in sugar-reduction strategies.

6.2.1 Nutritive (caloric) sweeteners

Nutritive sweeteners contribute energy and are used in gram-level quantities. They include sucrose, glucose, fructose, honey, jaggery, and syrups. Although some traditional sweeteners contain trace minerals or phytochemicals, they remain energy dense and contribute to free sugar intake (Saraiva et al., 2020). Therefore, they are not the primary focus for sugar reduction, except in partial replacement strategies (Tables 6.1, 6.2, and 6.3).

6.2.2 Low-calorie Sweeteners (bulk sweeteners/polyols)

Polyols (sugar alcohols) provide reduced energy (typically 0.2–2.6 kcal /g) and a lower glycemic response than sucrose. Importantly, polyols provide bulk and mouthfeel—making them highly valuable in nutraceutical formulations where sugar reduction is required but texture must be maintained (Magnuson et al., 2016).

6.2.3 Nonnutritive (high-intensity) sweeteners

High-intensity sweeteners (HIS) provide strong sweetness at very low concentrations and therefore contribute negligible calories. In nutraceutical systems, the most relevant HIS are natural sweeteners such as steviol glycosides (stevia) and mogrosides (monk fruit). HIS are essential tools for sugar reduction but cannot replace bulk and may require blending and bulking agents (O'Donnell and Kearsley, 2012; Saraiva et al., 2020).

6.3 Key challenges in sugar replacement

Successful sugar reduction depends on managing both sensory and technological consequences of removing sucrose. The most critical challenges are summarized below.

Table 6.1 Classification of sweeteners for nutraceutical food formulation.

Category	Energy contribution	Key examples	Relevance to nutraceutical foods
Nutritive sweeteners	~4 kcal /g	Sucrose, honey, syrups	Limited use (partial replacement)
Polyols (bulk sweeteners)	0.2–2.6 kcal /g	Erythritol, xylitol	Bulk + low-glycemic response
High-intensity sweeteners	~0	Stevia, monk fruit	Sweetness, no bulk; requires blends

6.3.1 Bulk and texture loss

Sucrose is a bulk ingredient contributing to volume, viscosity, and mouthfeel. In bakery products, sucrose influences crumb softness and structure. High-intensity sweeteners cannot replace this bulk because they are used in micro quantities; hence, bulking agents such as polyols, soluble fibers, or hydrocolloids are required (O'Donnell and Kearsley, 2012).

6.3.2 Aftertaste and temporal sweetness profile

Many natural high-intensity sweeteners exhibit delayed onset and lingering aftertaste compared with sucrose. Steviol glycosides may cause bitterness or licorice-like notes depending on glycoside composition. Therefore, sweetener blending (e.g., stevia + erythritol) is widely used to improve temporal profile and consumer acceptability (Saraiva et al., 2020).

6.3.3 Browning and flavor development

Sugar reduction often results in poor browning and reduced "baked" flavor because high-intensity sweeteners and many polyols do not support browning reactions. Rare sugars such as allulose can partially restore browning while supporting low-glycemic formulation goals (FDA, 2019; O'Donnell and Kearsley, 2012).

6.3.4 Shelf-life and stability

Sucrose contributes to shelf life through water binding and reduction of water activity. When replaced with HIS, additional humectants (polyols or fibers) may be required to maintain softness and microbial stability (O'Donnell and Kearsley, 2012).

6.4 Major nutraceutical sweeteners

Modern nutraceutical and sugar-reduced formulations rely primarily on four categories of sweeteners that balance sweetness delivery with technological

Table 6.2 Comparative profile of major nutraceutical sweeteners.

Sweetener	Class	Relative sweetness (vs. sucrose)	Calories	Key advantage	Main limitation
Steviol glycosides	Natural HIS	200–400×	~0	Zero glycemic impact, stable	Bitterness/aftertaste
Monk fruit mogrosides	Natural HIS	150–250×	~0	Clean sweetness, good for beverages	Cost, often needs blending
Erythritol	Polyol bulk sweetener	0.6–0.8×	~0.2 kcal / g	Bulk + high tolerance	Cooling effect
Xylitol	Polyol bulk sweetener	~1×	~2.4 kcal / g	Dental protection	Laxation at high intake
D-Allulose	Rare sugar	~0.7×	~0.4 kcal / g	Browning + bulk + low glycemic	Regulatory status evolving in some regions

Table 6.3 Summary of nutraceutical benefits of key sweeteners.

Sweetener	Glycemic response	Energy contribution	Dental impact	Notes for nutraceutical positioning
Stevia (steviol glycosides)	Minimal	~0	Noncariogenic	Useful for diabetes-focused products; blending improves sensory profile
Monk fruit (mogrosides)	Minimal	~0	Noncariogenic	Clean-label HIS; often blended
Erythritol	Negligible	Very low	Noncariogenic	Bulk support; cooling sensation
Xylitol	Low	Moderate	Strong anticariogenic	GI tolerance must be considered
Allulose	Minimal	Low	Neutral	Rare sugar with sugar-like functionality and browning

performance: (i) steviol glycosides (stevia), (ii) monk fruit mogrosides, (iii) polyols (bulk sweeteners), and (iv) rare sugars (allulose). These ingredients are selected not only for sweetness but also for their metabolic effects, sensory profile, stability during processing, and ability to meet clean-label expectations (O'Donnell and Kearsley, 2012; Saraiva et al., 2020).

6.4.1 Steviol glycosides (stevia)

6.4.1.1 Source and chemistry

Stevia is derived from the leaves of *Stevia rebaudiana* and contains a group of diterpene glycosides known as steviol glycosides. The most widely used compounds include stevioside and rebaudioside A, while newer purified fractions such as rebaudioside D and rebaudioside M have gained commercial importance due to improved sensory properties (JECFA, 2010; Saraiva et al., 2020).

6.4.1.2 Sweetness potency and sensory characteristics

Steviol glycosides are approximately 200–400 times sweeter than sucrose. Their main limitation is bitterness or licorice-like aftertaste, particularly at higher concentrations. For this reason, stevia is commonly used as part of blended sweetener systems (e.g., stevia + erythritol) to provide a sucrose-like sweetness profile (O'Donnell and Kearsley, 2012; Saraiva et al., 2020).

6.4.1.3 Stability and processing suitability

Steviol glycosides are relatively stable across food processing conditions, including thermal stability under typical cooking temperatures and good stability across a wide pH range. This makes stevia suitable for beverages, dairy products, tabletop sweeteners, and bakery formulations (with bulking support) (JECFA, 2010).

6.4.1.4 Physiological relevance

Stevia provides sweetness with negligible energy contribution and minimal glycemic effect. This is a major advantage in products formulated for diabetes management and calorie control (Magnuson et al., 2016).

6.4.2 Monk fruit (mogrosides)

6.4.2.1 Source and chemistry

Monk fruit (*Siraitia grosvenorii*) is a traditional sweetener source, with sweetness primarily attributed to mogrosides, especially mogroside V. These are triterpene glycosides that provide high-intensity sweetness with low caloric contribution (Saraiva et al., 2020).

6.4.2.2 Sweetness potency and sensory characteristics

Monk fruit extracts are typically 150–250 times sweeter than sucrose. They are often perceived as having a cleaner sweetness than early-generation stevia products. However, they are still commonly blended with other sweeteners to improve mouthfeel and sweetness timing (Saraiva et al., 2020).

6.4.2.3 Stability and processing suitability

Monk fruit sweeteners show good stability in acidic conditions and are widely used in beverages and nutraceutical drinks. They also perform well under moderate heat processing, though complete thermal stability depends on extract composition (Saraiva et al., 2020).

6.4.2.4 Nutraceutical relevance

Beyond sweetening, mogrosides have been explored for antioxidant activity in experimental settings. In nutraceutical foods, monk fruit is mainly valued for zero-glycemic sweetness with clean-label appeal (Saraiva et al., 2020).

6.4.3 Polyols (bulk sweeteners): erythritol and xylitol

Polyols are essential in sugar reduction because they provide both sweetness and bulk. Unlike high-intensity sweeteners, polyols can replace part of the physical structure normally provided by sucrose (Magnuson et al., 2016).

6.4.3.1 Erythritol

Erythritol is a four-carbon polyol produced commercially through fermentation. It provides very low energy (approximately 0.2 kcal /g) and has high digestive tolerance because it is largely absorbed and excreted unchanged (Magnuson et al., 2016).

A key sensory characteristic of erythritol is a cooling mouthfeel due to endothermic dissolution (negative heat of solution). This effect is useful in mints and gums but may require balancing in bakery and dairy formulations (O'Donnell and Kearsley, 2012).

6.4.3.2 Xylitol

Xylitol has sweetness close to sucrose and is well known for anticariogenic effects. It is not fermented by cariogenic oral bacteria and therefore reduces acid production and dental caries risk. It is widely used in chewing gum, oral-health foods, and sugar-free confectionery (EFSA, 2015; Magnuson et al., 2016).

6.4.3.3 Tolerance and labeling

Polyols are incompletely absorbed, and excessive intake may cause gastrointestinal discomfort such as bloating or laxation. Therefore, regulatory authorities often require advisory labeling for high-polyol foods (EFSA, 2015).

6.5 Rare sugars: D-allulose (a high-value nutraceutical sweetener)

6.5.1 Definition and uniqueness

Rare sugars are monosaccharides found in nature at low levels but increasingly produced commercially via enzymatic conversion. Among these, D-allulose (also called D-psicose) has become highly important for modern reduced-sugar food design because it provides sugar-like functionality with minimal energy contribution (FDA, 2019).

6.5.2 Functional advantages

Unlike most high-intensity sweeteners, allulose can contribute to:

- Bulk (partial replacement)
- Browning development
- Mouthfeel improvement
- Improved sensory similarity to sucrose

These properties make allulose highly suitable for bakery, confectionery, and dairy formulations where structure is crucial (O'Donnell and Kearsley, 2012; FDA, 2019).

6.5.3 Physiological relevance

Allulose exhibits minimal effect on blood glucose and provides low caloric contribution due to limited metabolism. It is therefore considered a promising ingredient for metabolic health-oriented foods (FDA, 2019).

Relative sweetness values vary by matrix and temperature; listed values are typical ranges used in formulation science. (O'Donnell and Kearsley, 2012; Magnuson et al., 2016)

6.6 Nutraceutical and physiological benefits of sweetener substitution

The nutraceutical relevance of sweeteners lies primarily in their ability to reduce dietary free sugars while supporting metabolic health outcomes. Sugar replacement using nonsugar sweeteners can reduce energy intake and glycemic load, which is particularly important in populations with obesity and type 2 diabetes. However, it is essential to interpret "health benefits" carefully, as long-term outcomes depend on total dietary quality and behavioral factors, and not only on sweetener choice (WHO, 2015; WHO, 2023).

6.6.1 Energy reduction and weight-related outcomes

Replacing nutritive sugars with nonsugar sweeteners can reduce energy density in foods and beverages. This is widely applied in reduced-calorie beverages and diet products. However, WHO has emphasized that while nonsugar sweeteners (NSS) may support short-term reduction of free sugar intake, evidence does not consistently show long-term benefits for sustained weight management, and some observational evidence suggests possible associations with adverse metabolic outcomes (WHO, 2023).

Nevertheless, regulatory bodies continue to permit approved nonsugar sweeteners within established safety limits. Thus, the application of nutraceutical sweeteners should focus on sugar reduction and product reformulation rather than substitution without dietary behavior change

Thus, nutraceutical sweeteners should be positioned primarily as:

- Tools for reducing free sugars, and
- Enabling reformulation, rather than as independent weight-loss agents.

6.6.2 Glycemic control and diabetes management

One of the strongest physiological advantages of natural high-intensity sweeteners (stevia, monk fruit) and several low-glycemic alternatives (erythritol, allulose) is minimal effect on postprandial blood glucose and insulin response.

6.6.2.1 Stevia

Steviol glycosides provide sweetness without increasing glycemic load and are therefore valuable in diabetic-friendly foods. Mechanistic evidence suggests potential influence on glucose homeostasis, but the nutraceutical claim should remain conservative unless supported by strong human clinical evidence in the target population (Magnuson et al., 2016; Saraiva et al., 2020).

6.6.2.2 Erythritol and xylitol

Polyols generally show low-glycemic responses. Erythritol is absorbed and excreted unchanged, contributing minimal energy and negligible glycemic impact. Xylitol has low-glycemic impact compared with sucrose, supporting its use in diabetic-friendly confectionery (Magnuson et al., 2016).

6.6.2.3 Allulose

Allulose is a high-value rare sugar because it provides sugar-like sensory and functional properties with limited metabolism and reduced caloric value. It is increasingly considered a key ingredient for reduced-sugar bakery and confectionery while maintaining acceptable texture and browning (FDA, 2019).

6.6.3 Dental health benefits

Dental caries is strongly driven by frequent intake of fermentable sugars, which are metabolized by oral bacteria to produce acids that demineralize tooth enamel. Sugar replacement is therefore a nutritionally relevant and evidence-based strategy for caries prevention.

6.6.3.1 Xylitol

Xylitol has the most clearly established dental advantage among sweeteners. It cannot be efficiently metabolized by cariogenic bacteria such as Streptococcus mutans, reducing acid production and supporting oral health. Consequently, xylitol is widely used in chewing gum and oral-care formulations (EFSA, 2015; Magnuson et al., 2016).

6.6.3.2 High-intensity sweeteners

Natural HIS (stevia, monk fruit) and synthetic HIS are generally nonfermentable and therefore noncariogenic. Their inclusion in sugar-free foods supports caries prevention strategies by reducing sucrose exposure frequency (WHO, 2015).

6.6.4 Gut microbiota considerations

The gut microbiota has a central role in host metabolism, immune regulation, and inflammation. Sweeteners may influence gut microbiota through different mechanisms:

6.6.4.1 Polyols

Some polyols are partially fermented in the colon, contributing to the production of short-chain fatty acids (SCFAs). This fermentation can support gut physiology, but higher intake may cause gastrointestinal symptoms due to osmotic effects and fermentation gas production (Magnuson et al., 2016; EFSA, 2015).

6.6.4.2 Nonsugar sweeteners

WHO's 2023 guideline highlights that evidence on NSS and long-term metabolic outcomes remains mixed and evolving. Therefore, microbiome-related nutraceutical claims should not be overstated and should be communicated as "under investigation" rather than confirmed benefits (WHO, 2023).

6.6.5 Antioxidant and bioactive potential

Among sweeteners, bioactive antioxidant contribution is most evident in honey and some plant-derived extracts; however, honey remains a caloric sweetener and should be framed as a "*functional nutritive sweetener*" rather than a sugar substitute for diabetes.

Monk fruit mogrosides have been explored experimentally for antioxidant potential, supporting nutraceutical positioning, but in foods their principal value remains clean sweetness with low/no glycemic impact (Saraiva et al., 2020; WHO, 2015; Magnuson et al., 2016; EFSA, 2015; FDA, 2019).

6.7 Safety and regulation

Safety evaluation of sweeteners is governed by internationally accepted risk-assessment principles, including toxicological evaluation, metabolic fate, dietary exposure estimation, and establishment of safe intake limits where applicable. For many food additives, safety is expressed using the Acceptable Daily Intake (ADI), defined as the amount that can be consumed daily over a lifetime without appreciable risk (WHO, 2023).

Key safety points relevant to nutraceutical sweeteners

- Steviol glycosides (stevia) have been evaluated by international authorities and specifications/ADI values have been established; they are widely permitted in foods and beverages in many regions (JECFA, 2010).
- Monk fruit extracts are increasingly used in clean-label products; regulatory status varies by region, with acceptance in some markets based on safety evaluations and historical use (Saraiva et al., 2020).
- Polyols (erythritol, xylitol, sorbitol, maltitol) are generally recognized as safe within intended use levels; however, high intake may cause gastrointestinal discomfort due to incomplete absorption, hence advisory labeling in several jurisdictions (EFSA, 2015; Magnuson et al., 2016).
- Rare sugars such as D-allulose have gained regulatory acceptance in some markets; labeling and permitted uses continue to evolve, requiring confirmation using current national standards during product development (FDA, 2019).

In 2023, WHO issued guidance advising against the routine use of nonsugar sweeteners for long-term weight control, emphasizing that NSS should support reduced free sugar intake but should not replace improvements in overall dietary quality (WHO, 2023). Importantly, this guideline does not imply that approved sweeteners are unsafe within regulatory limits, but highlights uncertainty regarding long-term population health outcomes.

6.8 Conclusion

Nutraceutical sweeteners represent a key innovation in modern food science, enabling sugar reduction while supporting metabolic health-oriented formulation. This chapter classified sweeteners based on metabolic contribution and emphasized the major technological challenges encountered in sugar replacement, including loss of bulk, altered sweetness temporal profile, reduced browning, and changes in shelf stability.

Among available options, steviol glycosides and monk fruit mogrosides serve as clean-label high-intensity sweeteners suitable for low-glycemic and calorie-reduced foods, while polyols (erythritol, xylitol) provide essential bulk and functionality in sugar-reduced products. D-allulose, as a rare sugar, offers a high-value alternative due to sugar-like functional performance including browning and (EFSA, 2015; FDA, 2019; JECFA, 2010; Magnuson et al., 2016; O'Donnell and Kearsley, 2012; Saraiva et al., 2020; WHO, 2015; WHO, 2023) mouthfeel with minimal glycemic impact.

Overall, the best nutraceutical strategy is not single-sweetener substitution, but formulation using sweetener systems that combine high-intensity sweeteners for sweetness with bulk sweeteners and functional ingredients for texture and stability. Evidence-based use of nutraceutical sweeteners can support reduced free sugar consumption and product innovation, provided that claims remain scientifically substantiated and aligned with regulatory requirements.

AI disclosure

During the preparation of this work, the authors used Open AI to correct grammar and for spell check. After using this tool/service, the authors reviewed and edited the content as needed and take full responsibility for the content of the publication.

References

EFSA (European Food Safety Authority), 2015. Scientific opinions and safety evaluations relevant to polyols (xylitol, sorbitol, erythritol) used as food additives/sweeteners. EFSA J.

FDA (U.S. Food and Drug Administration), 2019. Regulatory and Labeling Considerations Relevant to D-allulose (Rare Sugar) and GRAS/Labeling Updates. U.S. Food and Drug Administration, Silver Spring, MD.

JECFA (Joint FAO/WHO Expert Committee on Food Additives), 2010. Steviol Glycosides: Safety Evaluation and Specifications. FAO/WHO JECFA Monographs, Geneva.

Magnuson, B.A., Carakostas, M.C., Moore, N.H., Poulos, S.P., Renwick, A.G., 2016. Biological fate of low-calorie sweeteners. Nutrition Reviews 74 (11), 670–689. Available at https://doi.org/10.1093/nutrit/nuw032, http://nutritionreviews.oxfordjournals.org/.

O'Donnell, K., Kearsley, M.W., 2012. Sweeteners and Sugar Alternatives in Food Technology. Wiley. Available at http://doi.org/10.1002/9781118373941.

Saraiva, A., Carrascosa, C., Raheem, D., Ramos, F., Raposo, A., 2020. Natural sweeteners: The relevance of food naturalness for consumers, food security aspects, sustainability and health impacts. International Journal of Environmental Research and Public Health 17 (17), 6285. Available at https://doi.org/10.3390/ijerph17176285.

WHO (World Health Organization), 2015. Guideline: Sugars Intake for Adults and Children. World Health Organization, Geneva.

WHO (World Health Organization), 2023. Use of Non-sugar Sweeteners: WHO Guideline. World Health Organization, Geneva.

Natural flavors and flavor enhancers

7

Shilpa P.[1] *and B.K. Indu*[2]
[1]Department of Nutrition and Dietetics, Padmashree Institute of Management and Sciences, Bengaluru, Karnataka, India, [2]Department of Biotechnology, Padmashree Institute of Management and Sciences, Bengaluru, Karnataka, India

7.1 Natural flavors introduction

Flavor is one of the most important characteristics of food, as it largely determines whether a food is liked or disliked by consumers. Natural flavors are obtained from plant, animal, or microbial sources using physical, enzymatic, or traditional processing methods and are considered safer and more acceptable than synthetic flavors (Reshna et al., 2022). It strongly influences food choice, appetite, and overall eating satisfaction. In recent years, there has been a growing preference for natural flavors due to increased consumer awareness about health, food safety, and clean-label products (Hossain MS et al., 2025). Natural flavors are complex mixtures of chemical compounds that provide characteristic taste and aroma to food. Unlike artificial flavors, which may consist of single synthesized compounds, natural flavors contain multiple volatile and nonvolatile components that together produce a balanced and authentic flavor profile (Ogori, 2022).

7.1.1 Concept of flavors

Flavor is not a single sensation but a combined perception of taste, aroma, and mouthfeel. It is experienced when food is eaten and involves the coordinated action of several sensory systems. Taste is perceived by taste buds on the tongue and includes five basic tastes: sweet, sour, salty, bitter, and umami (Reshna et al., 2022). Aroma is detected by the olfactory receptors in the nose through volatile compounds released during chewing and swallowing, a process known as retronasal olfaction. In addition to taste and aroma, texture and temperature also influence flavor perception. For example, the crispness of fruits or the creaminess of dairy products can enhance the overall flavor experience (Hossain MS et al., 2025). Because flavor involves multiple senses working together, it is often described as a sensory complex rather than a single property of food (Reshna et al., 2022). Natural flavor compounds are broadly classified into volatile and nonvolatile compounds. Volatile compounds mainly contribute to aroma, such as esters, aldehydes, terpenes, and sulfur compounds. Nonvolatile compounds mainly contribute to taste and include organic acids, phenolic compounds, and alkaloids. The type, concentration, and interaction of these compounds determine the characteristic flavor of foods such as fruits, vegetables, spices, beverages, and fermented products (Fig. 7.1).

Nutraceutical Applications in Food Processing and Preservation. DOI: https://doi.org/10.1016/B978-0-443-45784-5.00008-3

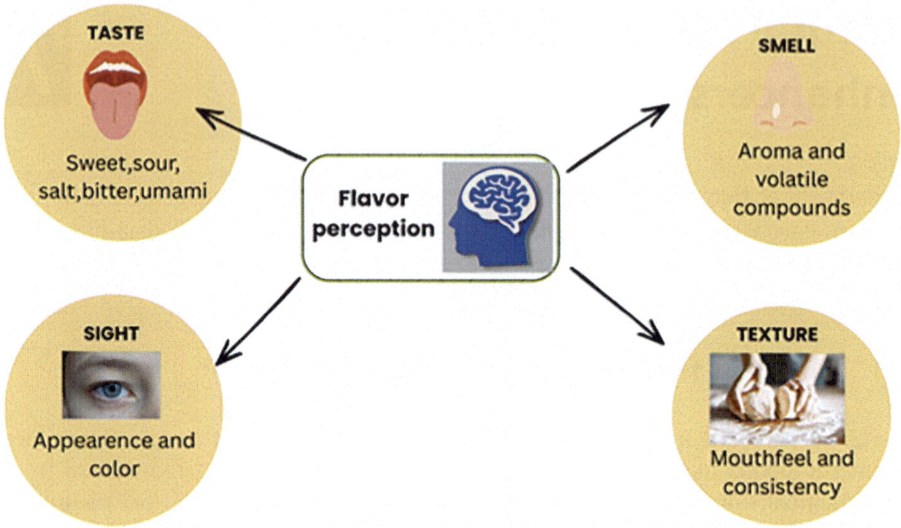

Figure 7.1 Multisensory factors influencing flavor perception. Integration of taste, smell, sight, and texture in the brain that collectively determines flavor perception and overall food sensory experience.

7.2 Role of flavors in foods

Beyond taste, flavors play several important roles in food

1. Improvement of palatability: Natural flavors enhance the taste and aroma of foods, making them more appealing and enjoyable. This increased palatability encourages food consumption and improves consumer acceptance of various food products.
2. Stimulation of appetite and digestion: Flavor compounds stimulate sensory receptors that promote salivation and the release of digestive juices. This helps prepare the body for digestion and improves nutrient utilization.
3. Masking of undesirable tastes: Natural flavors are often used to mask unpleasant tastes such as bitterness, metallic notes, or off-flavors that may arise during food processing, storage, or fortification.
4. Product identity and consumer preference: Flavors give foods their unique identity. For example, the characteristic aroma of coffee or the pungency of spices helps consumers recognize and prefer specific products. Small changes in flavor composition can significantly affect consumer choice.
5. Psychological and emotional influence: Flavors are closely linked to memory and emotions. Pleasant natural flavors can evoke positive feelings, comfort, and satisfaction, thereby enhancing overall well-being and food enjoyment.
6. Support for clean-label and natural foods: With increasing demand for natural and minimally processed foods, natural flavors support clean-label formulations. They also align with consumer expectations for safety, authenticity, and transparency in food products.

7.3 Importance of natural flavors

Natural flavors play a crucial role in food quality, sensory perception, and consumer behavior, making them essential components in food formulation and product development.

7.3.1 Sensory appeal

Natural flavors significantly enhance the sensory appeal of foods by improving taste and aroma, which are primary determinants of food enjoyment (Ogori, 2022). Flavor perception involves the integration of taste, aroma, and mouthfeel, creating a pleasurable eating experience that increases sensory satisfaction (Prescott, 2015). Pleasant natural flavor compounds stimulate olfactory and gustatory receptors, resulting in positive hedonic responses and improved overall sensory quality of foods (Hossain MS et al., 2025). Natural flavors often provide complex and authentic sensory profiles that are closer to real foods when compared to synthetic flavoring substances (Reshna et al., 2023).

7.3.2 Consumer acceptance

Consumer acceptance of food products is strongly influenced by flavor quality, as flavor directly affects liking, preference, and repeat consumption. Foods formulated with natural flavors are generally perceived by consumers as safer, healthier, and more trustworthy than those containing artificial flavoring agents (Asioli et al., 2017). The use of natural flavors supports clean-label trends and positively influences consumer willingness to purchase food products (Reshna et al., 2022). Studies have shown that ingredient information related to natural flavoring increases perceived product naturalness, which in turn enhances consumer acceptance (Román et al., 2017).

7.3.3 Product identity

Natural flavors play an important role in establishing and maintaining the unique identity of food products. Characteristic natural flavors help consumers recognize, differentiate, and associate specific sensory attributes with particular foods and brands. Consistency in natural flavor profiles is critical for maintaining brand identity and ensuring uniform product quality across different production batches (Asioli et al., 2017). Distinctive natural flavors also contribute to cultural familiarity and emotional attachment, strengthening consumer loyalty and long-term product preference.

7.4 Types and sources of natural flavors

Natural flavors are flavoring substances obtained from natural raw materials of plant, animal, or microbial origin, using physical, enzymatic, or microbiological processes, where the primary function is flavoring rather than nutrition (Reineccius, 2006).

Based on their biological origin, natural flavors are broadly classified into plant sources, animal sources, and microbial sources (Belitz, Grosch and Schieberle, 2009).

7.4.1 Plant sources

Plant sources constitute the largest and most diverse group of natural flavoring substances used in food products. Natural flavors from plants are obtained from various parts such as fruits, vegetables, leaves, seeds, bark, roots, flowers, herbs, and spices (Ogori, 2022). Fruit-derived flavors include citrus oils, berry extracts, and tropical fruit essences, which are rich in volatile compounds such as esters, aldehydes, and terpenes that impart characteristic fruity aromas (Reineccius, 2006). Vegetables such as onion and garlic contribute sulfur-containing compounds responsible for their pungent and savory flavor notes (Belitz, Grosch and Schieberle, 2009).

Herbs and spices such as mint, basil, cinnamon, clove, cardamom, and pepper provide complex aroma profiles due to the presence of essential oils and oleoresins. Plant-based natural flavors are widely used in beverages, bakery products, confectionery, dairy products, and savory foods because of their wide availability and consumer acceptance (Ogori, 2022).

7.4.2 Animal sources

Animal-derived natural flavors are obtained from edible animal tissues and products such as meat, poultry, seafood, dairy products, eggs, and fats (Reineccius, 2006). These flavors are particularly important in imparting savory, meaty, and umami characteristics to foods (Belitz, Grosch and Schieberle, 2009). Meat and poultry flavors are primarily generated through the Maillard reaction and lipid oxidation during cooking or controlled processing, producing compounds such as heterocyclics, aldehydes, and sulfur compounds (Belitz, Grosch and Schieberle, 2009). Dairy-based flavors such as butter, cheese, cream, and milk flavors arise from enzymatic and microbial breakdown of proteins and fats, leading to the formation of free fatty acids, ketones, and lactones. Seafood-derived flavors originate from amino acids, peptides, and lipid-derived compounds that contribute to characteristic marine and umami notes (Reineccius, 2006). Animal-derived natural flavors are commonly used in savory foods, ready-to-eat meals, soups, sauces, and meat analogs (Ogori, 2022).

7.4.3 Microbial sources

Microbial sources of natural flavors involve the use of microorganisms such as yeasts, bacteria, and fungi to produce flavor compounds through fermentation or biotransformation processes. Microbial flavor production is considered natural when the substrates and microorganisms used are of natural origin and the flavor compounds produced occur naturally (Reineccius 2006). Yeasts and lactic acid bacteria produce flavor compounds such as alcohols, esters, organic acids, and carbonyl compounds during fermentation of foods like bread, yogurt, cheese, and alcoholic beverages (Belitz, Grosch and Schieberle, 2009). Fungi such as *Aspergillus* and *Penicillium* species contribute to characteristic flavors in fermented foods and cheeses through the production of methyl ketones, alcohols, and other volatile metabolites. Microbial flavor production offers advantages such as consistency, sustainability, and the ability to produce specific flavor compounds under controlled conditions (Ogori, 2022). As a result, microbial sources are increasingly used for the production of bio-based natural flavors in the food industry (Reineccius, 2006).

7.5 Classification of natural flavors

Natural flavors can be classified based on the type of preparation and method of obtaining flavoring constituents from raw materials (plant, animal, microbial) (Reineccius, 2006). According to regulatory definitions and food science literature, natural flavor preparations include essential oils, oleoresins, extracts, and fermentation products (Britannica Editors, 2025). Each class differs in composition, sensory impact, and method of preparation.

7.5.1 Essential oils

Essential oils are highly concentrated volatile flavor compounds obtained mainly from plant materials through steam distillation or expression (Reineccius, 2006; Britannica Editors, 2025). They are composed of complex mixtures of organic compounds such as terpenes, aldehydes, ketones, alcohols, and esters that contribute strongly to characteristic food aromas (Britannica Editors, 2025). Examples include lemon oil from citrus peel, peppermint oil from mentha leaves, and clove oil from *Syzygium aromaticum* buds, which are used extensively in beverages, confectionery, baked products, and seasonings (Britannica Editors, 2025). Because essential oils contain predominantly volatile flavor constituents, they are particularly effective in imparting intense aroma but often lack the nonvolatile taste components of the original material. Essential oils are typically used at very low levels due to their high potency, and their stability can be enhanced through methods such as microencapsulation when incorporated into food matrices (Fernandes et al., 2024).

7.5.2 Oleoresins

Oleoresins are concentrated liquid extracts that contain both the volatile essential oil fraction and nonvolatile resinous components of the source material. They are usually produced by solvent extraction of spices, herbs, or other plant tissues, followed by solvent removal. For example, black pepper oleoresin contains piperine and other nonvolatiles along with volatile aroma compounds, while ginger oleoresin contains pungent compounds such as gingerol, shogaol, and zingerone. Because oleoresins represent a more complete flavor profile (both aroma and taste components), they are widely used in spice blends, sauces, soups, meat products, and snacks where a full flavor character is desired. Oleoresins are less volatile and often more stable than essential oils, making them suitable for complex flavor development in processed foods.

7.5.3 Extracts

Extracts are flavor preparations obtained by solvent extraction, maceration, or distillation of food materials, wherein the flavor constituents are separated from the raw material into a solvent matrix (Britannica Editors, 2025). These extractions can be alcoholic, aqueous, or use other food-grade solvents such as glycerol and propylene glycol depending on the desired flavor profile and application (Britannica Editors, 2025). Extracts may be derived from fruits (e.g., vanilla, berry extracts), herbs (e.g., basil, oregano), or plant spices, capturing both volatile and some semi-volatile constituents (Britannica Editors, 2025). Unlike essential oils, extracts often contain solvent-soluble compounds, which can provide a taste component as well as aroma, and are used widely in beverages, dairy products, bakery items, and confectionery. Extracts are sometimes categorized further into true extracts and rectified or folded oils, where unwanted components are removed or concentrated to improve flavor quality (MedCrave, 2025).

7.5.4 Fermentation products

Fermentation products as natural flavors are generated when microorganisms such as yeasts, bacteria, or fungi act on substrates to produce flavor compounds through metabolic activities (Reineccius, 2006;). These processes can involve enzymatic biotransformation of sugars, amino acids, and lipids into flavor compounds such as esters, acids, alcohols, aldehydes, and other volatiles that are naturally found in foods (Reineccius, 2006;). Common examples include cheese flavors developed by lactic acid bacteria and molds, yeast-derived fruity esters in fermented beverages, and buttery diacetyl in cultured dairy products, all of which contribute complex sensory notes to foods. Because fermentation is a biological process, it can produce flavors that are more complex and characteristic than those obtained by simple extraction alone. Fermentation products are considered natural if the substrate and microbial strains are of natural origin and the flavor compounds produced are found in nature or traditional foods, aligning with regulatory definitions of natural flavors

7.6 Natural versus synthetic flavors

Natural flavors are obtained directly from plants, animals, or microbial fermentation products using physical processes such as extraction, distillation, pressing, or enzymatic treatment. According to the American Chemical Society, a natural flavor is defined as one that is "derived from plant or animal source materials and isolated by physical methods without altering the chemical nature of the flavoring constituents". These flavoring substances include essential oils, oleoresins, and extracts obtained from fruits, vegetables, spices, herbs, and dairy products.

7.6.1 Examples of natural flavors

7.6.1.1 Vanilla extract

Vanilla flavor is naturally obtained from the cured pods of *Vanilla planifolia*. The characteristic flavor arises from a complex mixture of compounds, predominantly vanillin, along with minor phenolic and aromatic constituents. The extraction is carried out using aqueous alcohol, producing a flavor profile that is rich and multidimensional compared to synthetic vanillin (Britannica Editors., 2025).

7.6.1.2 Citrus essential oils

Natural citrus flavors such as orange and lemon oil are obtained by cold pressing or steam distillation of citrus peels. These oils contain volatile compounds like limonene, citral, and linalool, which provide fresh and characteristic citrus aromas widely used in beverages and confectionery (Reineccius, 2006).

7.6.1.3 Dairy-derived flavors (butter and cheese notes)

Natural butter and cheese flavors may be produced from dairy sources or through microbial fermentation, yielding compounds such as diacetyl and short-chain fatty acids. These flavors contribute to the authentic creamy and savory notes in bakery and dairy products (Reineccius, 2006).

In contrast, synthetic flavors are produced through chemical synthesis in laboratories to replicate or enhance the sensory properties of natural flavor compounds. According to the (Food Safety Institute., 2023), synthetic flavors may be either nature-identical—chemically identical to compounds found in nature—or artificial, meaning they do not occur naturally but are designed to produce a specific flavor effect. These flavors offer consistency, scalability, and cost efficiency for food manufacturers.

7.6.2 Examples of synthetic flavors

7.6.2.1 Nature-identical vanillin

Synthetic vanillin is chemically identical to natural vanillin but is produced from sources such as guaiacol or lignin through chemical synthesis. While it provides a strong vanilla note, it lacks the complex secondary compounds present in natural vanilla extract, resulting in a simpler flavor profile.

7.6.2.2 Ethyl butyrate (pineapple/orange flavor)

Ethyl butyrate is a synthetic ester commonly used to impart pineapple, orange, and tropical fruit flavors in beverages and confectionery. Although it occurs naturally in small quantities in fruits, the commercially used compound is produced synthetically to ensure uniformity and availability (Reineccius, 2006).

7.6.2.3 *Artificial strawberry flavor*

Synthetic strawberry flavor is typically created using a blend of esters such as ethyl methyl phenylglycidate and ethyl butyrate. These compounds mimic the sensory impression of strawberry despite the absence of many of the hundreds of aroma compounds found in real strawberries (Reineccius, 2006).

7.7 Safety and regulatory evaluation of natural and synthetic flavors

From a regulatory and toxicological perspective, the safety of flavoring substances is determined by their chemical identity, purity, metabolic fate, and level of dietary exposure, rather than whether they are classified as natural or synthetic (Reineccius, 2006). International food safety authorities evaluate flavoring agents based on toxicological data, acceptable daily intake (ADI), and intended use levels, applying the same scientific principles to both natural and synthetic compounds (Britannica Editors., 2025). Natural flavors are often perceived by consumers as inherently safer due to their biological origin. However, natural flavorings are typically complex mixtures of dozens to hundreds of compounds, whose composition can vary depending on plant variety, harvesting conditions, extraction methods, and storage (What to Know about Food Flavor: A Review, 2023). This variability can complicate safety evaluation, as minor constituents may exert biological effects at higher exposure levels.

Synthetic flavors, particularly nature-identical compounds, are chemically well-defined single molecules, which allows for precise toxicological assessment, reproducibility, and controlled exposure evaluation (Food Safety Institute., 2023). Regulatory agencies such as the FDA, EFSA, and JECFA assess both categories using standardized toxicological tests including acute, subchronic, chronic toxicity, genotoxicity, carcinogenicity, and reproductive toxicity studies (Reineccius, 2006). Importantly, regulatory approval does not rely on origin but on risk assessment, where hazard identification and exposure assessment are combined to determine safety margins (Britannica Editors., 2025).

Natural flavor compounds may still pose toxicological concerns when consumed in excess. For example, safrole, a naturally occurring compound found in sassafras oil, was restricted after studies demonstrated carcinogenic potential in animal models. This highlights that natural origin does not guarantee safety. Synthetic flavors have been scrutinized extensively due to their widespread use. While many are safe within regulatory limits, some compounds have been associated with respiratory, neurological, or gastrointestinal effects, particularly at occupational or high exposure levels (Reineccius, 2006). Consequently, regulatory agencies continuously reevaluate flavoring substances as new toxicological data emerge (Table 7.1).

Table 7.1 Evidence based adverse health effects of selected synthetic flavoring substances.

Synthetic flavoring substance	Use in foods	Reported adverse health effects	Supporting evidence (published studies)
Diacetyl (2,3-butane-dione)	Butter flavor in bakery, popcorn, dairy	Bronchiolitis obliterans ("popcorn lung") in occupational exposure; respiratory irritation	Hubbs et al. (2008)
Ethyl maltol	Sweet, caramel flavor in confectionery	High doses linked to oxidative stress and liver changes in animal models	Kim SA et al. (2018)
Vanillin (synthetic)	Vanilla flavor in beverages and bakery	Generally safe at low doses; high concentrations may cause headaches and allergic reactions in sensitive individuals	Walton et al. (2003)
Coumarin (synthetic)	Cinnamon-like flavor	Hepatotoxicity observed in animal studies; intake limits imposed	Lake (1999)

7.8 Sensory quality of natural and synthetic flavors

The sensory quality of natural and synthetic flavors reflects how humans perceive flavor through the interaction (Vasilaki et al., 2021) of volatile compounds with olfactory and gustatory receptors, the complex interactions among flavor compounds, and their release dynamics within food matrices. Contemporary food science research demonstrates that flavor perception is not merely the additive effect of individual aroma compounds, but rather arises from complex combinatorial interactions and dynamic perception patterns. This understanding helps explain why natural and synthetic flavors are often experienced differently during sensory evaluation (De Santis, 2024; Parker et al., 2025).

7.9 Natural flavors: complexity and multidimensional sensory profiles

Natural flavors are composed of hundreds of volatile and nonvolatile compounds, each contributing distinct sensory attributes such as fruity, floral, spicy, woody, or earthy notes. Research on plant-derived flavor compounds indicates that these complex mixtures interact synergistically with taste and olfactory receptors, enhancing sensory nuance and depth beyond what individual compounds can produce alone (Czerny et al., 2023).

Advanced analytical techniques such as gas chromatography–mass spectrometry (GC-MS) and gas chromatography–olfactometry (GC-O) have shown that natural flavor extracts deliver distinct top, middle, and base notes that evolve temporally during mastication and swallowing. This layered release pattern contributes to a multidimensional sensory experience perceived as richness and authenticity (De Santis, 2024). For example, citrus essential oils contain limonene as the dominant compound, but their characteristic freshness and balance arise from minor terpenes, aldehydes, and oxygenated compounds that impart green, floral, bitter, or sweet undertones (Chen et al., 2024). Because the composition of natural flavors varies with plant variety, maturity, geographical origin, and processing conditions, their sensory profiles may differ slightly between batches. Sensory panels often perceive this variability positively, associating it with naturalness, depth, and complexity, particularly in premium and minimally processed foods.

7.10 Synthetic flavors: precision, consistency, and targeted sensory notes

Synthetic flavors are generally formulated to emphasize specific sensory targets, such as fruity, buttery, or caramel notes, using a limited number of chemically well-defined compounds. These molecules are selected to activate specific sensory receptors, allowing precise control over flavor intensity and profile (Food Safety Institute., 2023). A major advantage of synthetic flavors is their high sensory consistency. Because they are produced from standardized chemical formulations, they deliver uniform flavor intensity and character across batches, unlike natural extracts that may vary due to raw material fluctuations. This reproducibility is particularly valuable in large-scale food manufacturing (Mortensen and Sørensen, 2023).

Sensory science studies also show that synthetic flavors often exhibit rapid onset and strong initial impact, followed by a shorter persistence, especially in aqueous or low-fat systems (Zhang et al., 2023). While synthetic flavors can closely replicate dominant aroma notes, they may lack subtle background nuances unless deliberately blended to simulate compound interactions. For instance, synthetic citrus flavors may reproduce the limonene-driven citrus note but miss minor green or floral nuances unless additional compounds are incorporated (Chen et al., 2024).

7.11 Perception and food matrix interactions

Flavor perception is strongly influenced by the food matrix, as volatile compounds interact with fats, proteins, and carbohydrates, affecting aroma release and retronasal perception. Studies by Boland et al. (2013) demonstrated that complex food matrices slow aroma release and extend flavor persistence, a phenomenon more commonly observed with natural flavor extracts due to their diverse chemical composition. In contrast, synthetic flavors, with simpler molecular profiles, often release more rapidly, producing intense but shorter-lasting sensory impressions. This difference in release dynamics contributes to distinct temporal sensory experiences between natural and synthetic flavors.

7.12 Consumer and industry perspectives

Consumer research consistently indicates that natural-flavored products are perceived as more authentic, fresh, and emotionally appealing, while synthetic flavors are valued for their clarity, stability, and consistency (Prescott, Young, 2024). From an industry perspective, natural flavors are preferred in premium, clean-label, and artisanal foods, whereas synthetic flavors are widely used in beverages, confectionery, and snack foods where cost efficiency, shelf stability, and uniform sensory impact are critical (Fernandes et al., 2024).

7.13 Cost and stability

Cost and stability are among the most critical factors influencing the choice between natural and synthetic flavors in food formulation. These factors affect not only product pricing but also processing feasibility, shelf life, and sensory consistency throughout distribution and storage.

7.13.1 Cost considerations

Natural flavors are generally more expensive due to their reliance on agricultural raw materials, which are subject to seasonal availability, climatic conditions, pest pressures, and geographical variability. The fluctuations in crop yield and quality can significantly influence both the availability and price of natural flavor sources such as vanilla beans, citrus peels, spices, and herbs. For example, natural vanilla is one of the most expensive flavoring agents due to limited growing regions, labor-intensive cultivation, and curing processes (Britannica Editors., 2025).

In addition, the extraction and processing of natural flavors including steam distillation, solvent extraction, supercritical CO_2 extraction, or enzymatic treatments are technically demanding and costly. These processes often yield relatively low concentrations of flavor compounds, further increasing production costs (Reineccius, 2006). Postextraction stabilization steps such as concentration, standardization, and encapsulation also add to manufacturing expenses (Fernandes et al., 2024).

In contrast, synthetic flavors offer significant economic advantages. They are produced via chemical synthesis using readily available precursors, allowing large-scale, continuous production independent of agricultural constraints. This results in lower production costs, predictable pricing, and reliable supply chains, which are particularly important for multinational food manufacturers (Food Safety Institute., 2023). Consequently, synthetic flavors are often preferred in cost-sensitive food categories such as soft drinks, confectionery, and mass-produced bakery products.

7.13.2 Stability challenges

Stability is a major limitation of many natural flavors. Natural flavor compounds especially terpenes, aldehydes, esters, and phenolic compounds are often chemically sensitive to heat, light, oxygen, and moisture. During food processing operations such as baking, extrusion, pasteurization, or sterilization, these compounds may undergo oxidation, volatilization, or degradation, leading to flavor loss or development of off-notes (Parker et al., 2025). Storage conditions further affect the stability of natural flavors. Exposure to oxygen and light can accelerate oxidative reactions, while interactions with food matrix components such as fats and proteins can alter flavor release and perception over time (Boland et al., 2013). As a result, maintaining consistent flavor quality in products containing natural flavors often requires advanced stabilization strategies such as microencapsulation, antioxidant addition, or modified atmosphere packaging (Fernandes et al., 2024).

Synthetic flavors generally exhibit greater chemical and physical stability due to their simpler, well-defined molecular structures. These compounds are designed to withstand a wide range of processing conditions, including high temperatures, varying pH levels, and prolonged storage, with minimal loss of sensory quality (Food Safety Institute., 2023). Their predictable behavior under processing stresses contributes to longer shelf life and reduced formulation risk. From an industrial perspective, the superior stability and lower cost of synthetic flavors make them especially suitable for high-volume processed foods, where uniformity, shelf stability, and cost control are essential. According to (Reineccius, 2006), synthetic flavors provide manufacturers with greater flexibility in formulation and fewer quality variations across production batches.

Nevertheless, increasing consumer demand for clean-label and natural products has driven innovation in stabilizing natural flavors, including encapsulation technologies and biotechnological production methods. Although these approaches improve stability, they often further increase costs, reinforcing the ongoing trade-off between natural authenticity and economic feasibility (Fernandes et al., 2024; Britannica Editors., 2025).

7.14 Nutraceutical benefits of natural flavors: role of specific flavor compounds

Natural flavors contain a wide spectrum of bioactive flavor compounds such as phenolic compounds, flavonoids, terpenes, aldehydes, ketones, and sulfur-containing compounds. These molecules, while primarily responsible for aroma and taste, also exert important nutraceutical effects, contributing to antioxidant defense, digestive health, antimicrobial protection, antiinflammatory responses, and metabolic regulation (Reineccius, 2006; Hossain MS et al., 2025).

7.14.1 Antioxidant activity of specific flavor compounds

Several flavor compounds exhibit strong antioxidant properties by scavenging reactive oxygen species and inhibiting lipid peroxidation. Eugenol, a major compound in clove flavor, shows potent free-radical scavenging activity and protects cellular membranes from oxidative damage (Shahidi and Ambigaipalan, 2015). Cinnamaldehyde, responsible for cinnamon flavor, acts as an antioxidant by inhibiting lipid oxidation and enhancing endogenous antioxidant enzyme activity (Srinivasan, 2014). Carnosic acid and carnosol, key flavor compounds in rosemary, stabilize food lipids and reduce oxidative stress in biological systems (Loussouarn et al., 2013). Citrus flavanones such as hesperidin and naringenin, associated with citrus flavors, contribute to antioxidant protection and vascular health (Shahidi and Ambigaipalan, 2015). These compounds enhance both food stability and human antioxidant status when consumed regularly through flavor-rich foods.

7.14.2 Digestive support and gastrointestinal health

Many flavor compounds stimulate digestive processes and improve gut comfort. Menthol, the characteristic compound of peppermint flavor, relaxes gastrointestinal smooth muscles and reduces symptoms of bloating and irritable bowel discomfort (McKay & Blumberg, 2006). Gingerols and shogaols, responsible for ginger flavor, stimulate gastric motility, enhance bile secretion, and reduce nausea (Srinivasan, 2014) Anethole, the principal flavor compound in fennel and anise, exhibits carminative effects by reducing gas formation and intestinal spasms (Aggarwal and Harikumar, 2009). Limonene, a dominant citrus flavor compound, promotes gastric emptying and supports liver detoxification pathways (Pérez-Jiménez et al., 2018). These effects explain the traditional use of spice-based flavors as digestive aids.

7.14.3 Antimicrobial properties of flavor compounds

Several natural flavor compounds possess strong antimicrobial activity, contributing to food safety and preservation. Thymol and carvacrol, found in thyme and oregano flavors, disrupt bacterial cell membranes and inhibit growth of *E. coli*, *Salmonella*, and *Listeria* species (Burt, 2004). Cinnamaldehyde inhibits microbial enzyme systems and interferes with bacterial energy metabolism (Burt, 2004). Allicin, a sulfur-containing compound responsible for garlic flavor, exhibits broad-spectrum antibacterial and antifungal activity (HossainMS et al., 2025). Citral, present in lemongrass and citrus flavors, shows antimicrobial effects against both Gram-positive and Gram-negative bacteria (Fernandes et al., 2024). These compounds are increasingly used in natural food preservation systems.

7.14.4 Antiinflammatory effects of flavor compounds

Chronic inflammation is modulated by several flavor-derived bioactives. Curcumin, the principal compound in turmeric flavor, suppresses inflammatory signaling pathways such as NF-κB and reduces expression of proinflammatory cytokines (Aggarwal and Harikumar, 2009). Gingerols, from ginger flavor, inhibit cyclooxygenase (COX) and lipoxygenase (LOX) enzymes involved in inflammation (Srinivasan, 2014). Eugenol exhibits antiinflammatory action by reducing prostaglandin synthesis (Shahidi and Ambigaipalan, 2015). Limonene reduces inflammatory markers and oxidative stress in metabolic tissues (Pérez-Jiménez et al., 2018).

7.14.5 Metabolic and cardioprotective effects

Certain flavor compounds influence glucose and lipid metabolism. Cinnamaldehyde improves insulin sensitivity and regulates postprandial glucose response (Srinivasan, 2014). Naringenin, a citrus flavor-associated flavonoid, improves lipid metabolism and reduces LDL oxidation (Shahidi and Ambigaipalan, 2015). Phenolic terpenes in rosemary flavor support endothelial function and reduce cardiovascular oxidative stress (Pérez-Jiménez et al., 2018).

7.14.6 Neuroprotective and psychophysiological effects

Flavor compounds can also affect neurological function and emotional well-being. Linalool, present in coriander and lavender flavors, exhibits anxiolytic and neuroprotective effects by modulating neurotransmitter activity (Spence, 2023). Limonene has been associated with mood enhancement and stress reduction through olfactory stimulation (Prescott, Young, 2024). These findings highlight the brain–flavor–health connection.

7.14.7 Modulation of gut microbiota

Polyphenol-rich flavor compounds influence gut microbial composition. Flavonoids and phenolic acids present in spice and citrus flavors promote beneficial gut bacteria while suppressing pathogenic species (Pérez-Jiménez et al., 2018). This microbiota modulation contributes indirectly to improved metabolic and immune health.

7.15 Flavor enhancers

7.15.1 Concept of flavor enhancement

Flavor enhancers are food ingredients that intensify or modify the existing taste and aroma of foods without imparting a characteristic flavor of their own. Unlike flavoring agents, which add a specific aroma or taste, flavor enhancers act by amplifying inherent sensory attributes already present in the food matrix (Vasilaki et al., 2021). Their primary function is to improve overall palatability, balance taste perception, and enhance flavor fullness even at very low concentrations. The concept of flavor enhancement is closely associated with the perception of umami and kokumi sensations. Umami, recognized as the fifth basic taste, is primarily elicited by free L-glutamate and certain 5′-ribonucleotides, while kokumi refers to sensory attributes such as mouthfulness, continuity, and thickness that enhance flavor richness without possessing a distinct taste (Yamamoto and Inui-Yamamoto, 2023). These properties make flavor enhancers especially valuable in processed foods, low-salt formulations, and plant-based products where flavor intensity may otherwise be diminished (Amaliah et al., 2024).

Historically, flavor enhancers have been used in traditional cuisines long before industrial food processing. Fermented foods, aged cheeses, dried mushrooms, fish sauces, and seaweeds naturally contain high levels of glutamate and nucleotides that enhance savory taste (Kurihara, 2009; Amaliah et al., 2024). With the expansion of processed food markets in the mid-20th century, purified compounds such as monosodium glutamate (MSG) and disodium inosinate were introduced to standardize and intensify flavor across large-scale food production systems (Maga, Yamaguchi, 1983).

In recent years, consumer demand for clean-label and natural ingredients has driven renewed interest in naturally derived flavor enhancers obtained from plants, fungi, dairy products, and fermentation processes. Natural flavor enhancers are increasingly explored as alternatives to synthetic additives, particularly in view of consumer perceptions linking natural ingredients with safety, sustainability, and health benefits (Fadimatou, Bonheur and Loh, 2024; Vasilaki et al., 2021).

7.16 Mechanism of action

The mechanism of action of flavor enhancers is rooted in their interaction with human taste receptors and sensory signaling pathways. Umami compounds such as glutamate activate specific receptors located on taste bud cells, primarily the T1R1/T1R3 heterodimer and metabotropic glutamate receptors (mGluRs), leading to intracellular signaling cascades that enhance taste perception (Yamamoto and Inui-Yamamoto, 2023; Zhang et al., 2008).

A defining feature of umami compounds is their synergistic interaction. While glutamate alone produces a mild savory taste, its combination with 5′-ribonucleotides such as inosine monophosphate (IMP) and guanosine monophosphate (GMP) dramatically increases umami intensity by lowering taste detection thresholds (Yamaguchi, 1991; Nelson et al., 2002). This synergism explains why foods combining plant and animal ingredients—such as tomatoes with cheese or mushrooms with meat—exhibit heightened savoriness.

Flavor enhancers also exert effects beyond taste receptor activation. Recent research demonstrates that glutamate facilitates the binding of kokumi compounds, such as γ-glutamyl peptides, to calcium-sensing receptors, thereby enhancing mouthfeel, continuity, and flavor persistence (Yamamoto and Inui-Yamamoto, 2023). These interactions result in a fuller and more satisfying sensory experience without increasing flavor concentration or sodium content.

Additionally, flavor enhancers influence flavor perception by interacting with the food matrix. Proteins, fats, and carbohydrates modulate the release of volatile compounds during mastication, altering retronasal aroma perception and prolonging flavor duration (Fadimatou, Bonheur and Loh, 2024). This matrix-dependent release is particularly important in complex foods such as soups, sauces, and meat analogs. From a nutritional perspective, the use of flavor enhancers enables sodium reduction strategies. Studies show that incorporating umami substances allows up to 30% reduction in added salt without compromising palatability, supporting public health initiatives aimed at reducing cardiovascular risk (Kurihara, 2009; Amaliah et al., 2024).

7.17 Importance of flavor enhancers

7.17.1 *Taste intensification*

One of the primary roles of flavor enhancers is taste intensification, which is achieved by amplifying existing sensory attributes of food rather than imparting a new flavor of their own. Flavor enhancers increase the overall flavor impact by strengthening taste signals at the receptor level and by improving flavor integration within the food matrix (Vasilaki et al., 2021). Umami-related compounds, particularly L-glutamate and 5′-ribonucleotides, exhibit strong synergistic effects that significantly increase perceived taste intensity even at low concentrations (Kurihara, 2009). This synergy lowers taste detection thresholds and enhances flavor persistence, leading to foods that are perceived as richer and more satisfying (Yamamoto and Inui-Yamamoto, 2023). Such enhancement allows manufacturers to achieve desirable sensory strength without increasing the level of flavoring substances.

Formulated natural and organic flavor enhancers have also been shown to improve mouthfeel, continuity of flavor, and aftertaste, producing sensory attributes described as "roundness" and "depth" (Fadimatou, Bonheur and Loh, 2024). These characteristics are particularly important in processed foods, plant-based products, and reformulated foods where flavor dilution or loss may occur during processing. Additionally, flavor enhancers improve aroma–taste integration by facilitating the interaction between volatile aroma compounds and taste perception during mastication and retronasal olfaction, thereby enhancing overall flavor perception and consumer acceptance (Vasilaki et al., 2021).

7.17.2 *Salt and sugar reduction*

An important nutritional significance of flavor enhancers lies in their ability to support salt and sugar reduction while maintaining sensory acceptability. High intake of sodium and added sugars is a major public health concern, and flavor enhancers offer a technological strategy to reduce these components without compromising taste quality. Umami substances such as monosodium glutamate, yeast extracts, fermented protein hydrolysates, and mushroom concentrates have been shown to compensate for reduced salt levels by enhancing savory perception (Kurihara, 2009; Amaliah et al., 2024). The inclusion of these compounds allows sodium reductions of approximately 30–40% in foods such as soups, sauces, snacks, and meat products while maintaining consumer liking (Vasilaki et al., 2021). Mechanistically, umami compounds stimulate taste receptors that interact with salt perception pathways, thereby enhancing saltiness intensity even when sodium levels are lowered (Yamamoto and Inui-Yamamoto, 2023). This sensory compensation effect has been consistently demonstrated in both model food systems and commercial food formulations (Vasilaki et al., 2021).

In sugar-reduced foods, flavor enhancers contribute indirectly by improving overall flavor balance and suppressing bitterness or metallic aftertastes associated with high-intensity sweeteners. Natural flavor enhancers derived from fermentation and botanical sources enhance flavor fullness and mouthfeel, enabling partial sugar reduction in beverages and bakery products without negative sensory impact (Fadimatou, Bonheur and Loh, 2024). Compared to traditional umami substances, natural umami ingredients (mushrooms, seaweeds, tomatoes, soybeans, tea, grains) demonstrate superior functional and nutritional benefits. When used as salt-reduction strategies, they not only enhance flavor perception but also contribute to improved textural properties, increased water-holding capacity (WHC), and elevated antioxidant potential (Ma et al., 2024). Furthermore, the use of natural flavor enhancers aligns with clean-label and health-oriented product development, as ingredients such as yeast extracts, seaweed derivatives, and fermented flavor bases provide effective taste enhancement with minimal contribution to sodium, sugar, or caloric content (Amaliah et al., 2024).

7.18 Types and sources of flavor enhancers

In addition to amino acids, nucleotides, and fermentation-derived compounds, a broader array of flavor enhancers are used in food systems to intensify taste perceptions particularly umami, kokumi, mouthfulness, and retronasal aroma integration. These enhancers act by interacting with taste receptors, enhancing the response to basic tastes without imparting a distinctive flavor of their own (Vasilaki et al., 2021).

7.18.1 Amino acids and their salts

Amino acids remain the primary class of flavor enhancers due to their role in umami taste perception. Glutamic acid and its salts (e.g., monosodium glutamate, MSG) are the most widely studied and used enhancers. MSG enhances savory taste by activating umami receptors and increasing perceived flavor richness without dominating the flavor profile (Kurihara, 2009; Vasilaki et al., 2021). Other amino acids such as glycine, alanine, and aspartate also contribute to taste modulation; glycine can enhance sweetness and balance sour or bitter notes, while aspartate exhibits a mild umami effect and contributes to savory taste enhancement (Yamaguchi and Ninomiya, 2000; Reineccius, 2006).

In addition to free amino acids, small peptides short chains of amino acids have been found to elicit umami or kokumi taste effects (Stone, Herbert, 2012). For example, umami peptides occur naturally in enzymatic hydrolysates and cooked meat broths; these peptides enhance savory perception through synergistic interactions with glutamate and ribonucleotides (Yamamoto and Inui-Yamamoto, 2023).

7.18.2 Nucleotides and their derivatives

Nucleotides are potent flavor enhancers that work synergistically with glutamate to produce intense umami taste. The most important flavor-active nucleotides are inosine monophosphate (IMP) and guanosine monophosphate (GMP), which are naturally present in meat, seafood, mushrooms, and yeast extracts (Yamaguchi and Ninomiya, 2000; Kurihara, 2009). These ribonucleotides themselves have minimal taste at low concentrations, but when combined with glutamate they dramatically enhance umami perception by lowering taste detection thresholds at the receptor level (Kurihara, 2009; Yamamoto and Inui-Yamamoto, 2023).

Other nucleotide salts such as disodium cytidylate (CMP) and disodium adenylate (AMP) are also used as flavor enhancers, although IMP and GMP exhibit the strongest synergistic effects (Vasilaki et al., 2021). Commercial blends of IMP and GMP (often labeled as E635 or "I+G") are widely used in savory foods because they provide enhanced umami with greater stability and potency than glutamate alone (Reineccius, 2006). Some calcium 5′-ribonucleotides also function as flavor enhancers by interacting with calcium-sensing receptors on taste buds, producing richer taste perception and enhanced flavor intensity without increasing salt content (Ohsu et al., 2010).

7.18.3 Fermentation-derived compounds

Many flavor enhancers originate from biological fermentation processes, which generate complex mixtures of taste-active substances including free amino acids, nucleotides, peptides, and organic acids. Yeast extracts obtained from controlled autolysis or enzymatic hydrolysis of *Saccharomyces cerevisiae* cells are rich sources of glutamate, nucleotides, and peptides that enhance umami and contribute to mouthfeel (Fadimatou, Bonheur and Loh, 2024; Vasilaki et al., 2021). These extracts are widely used in soups, sauces, processed meats, and snack foods. Hydrolyzed vegetable proteins (HVPs) are produced by breaking down plant proteins (e.g., soy, corn, wheat) into constituent amino acids and peptides. HVPs release glutamate and peptides that significantly enhance savory taste and are used in broths, gravies, and ready-to-eat meals (Reineccius, 2006; Vasilaki et al., 2021).

Other fermentation products, such as fermented soy sauce, fish sauce, and miso, contain natural flavor enhancers developed through microbial activity that generate glutamate, IMP, GMP, and short peptides, contributing to rich umami taste in traditional cuisines (Yamaguchi and Ninomiya, 2000).

7.18.4 Organic acids and miscellaneous taste modulators

Beyond the classical categories, several organic acids and Maillard reaction products act as secondary flavor enhancers by modulating taste perception. Organic acids such as succinic acid, lactic acid, and propionic acid contribute to mouthfulness and, in combination with umami compounds, deepen savory flavor profiles (Kurihara, 2009; Fadimatou, Bonheur and Loh, 2024). Recent analytical studies have identified thiamine-derived compounds formed during thermal processing as flavor modulators with kokumi-like properties, enhancing flavor continuity and richness without imparting a specific taste (Mottram, 2021).

Additionally, naturally occurring flavor precursors such as sulfur-containing amino acids (e.g., cysteine and methionine) participate in Maillard reactions during cooking, forming meaty and roasted flavor notes that enhance taste perception in complex food matrices, although these are not traditionally used as isolated flavor enhancers (Reineccius, 2006).

7.18.5 Combined and natural source enhancers

Many food ingredients are naturally rich in flavor-enhancing compounds and can be considered natural flavor enhancer sources. Seaweeds such as kombu and kelp contain high levels of free glutamate and ribonucleotides and have traditionally been used to enhance umami taste in soups and stocks (Kurihara, 2009). Tomatoes and mushrooms, particularly shiitake, are plant sources of glutamate and nucleotides that naturally boost savory taste in sauces, gravies, and plant-based foods (Yamaguchi and Ninomiya, 2000; Vasilaki et al., 2021). Cheese, especially aged varieties such as Parmesan, and other aged protein foods contain free amino acids and nucleotides that enhance savory perception and flavor depth (Reineccius, 2006). These natural ingredient sources are increasingly used in clean-label product development to deliver flavor enhancement without the addition of synthetic flavoring substances.

7.19 Natural versus synthetic flavor enhancers

Flavor enhancers may be classified as natural or synthetic based on their source and method of production. While both categories serve the same functional purpose enhancing taste perception their origin, regulatory considerations, and consumer acceptance differ significantly.

7.19.1 Source origin

Natural flavor enhancers are obtained from biological sources, including plants, animals, or microorganisms, using physical, enzymatic, or fermentation processes. Common examples include yeast extracts, fermented soy products, mushroom concentrates, seaweed extracts, and naturally occurring amino acids and nucleotides derived from protein-rich foods (Vasilaki et al., 2021). These enhancers contain complex mixtures of glutamate, ribonucleotides, peptides, and organic acids that act synergistically to enhance umami and mouthfulness. Synthetic flavor enhancers, in contrast, are produced through chemical synthesis or controlled industrial fermentation, resulting in highly purified and chemically defined compounds. Examples include monosodium glutamate (MSG), disodium inosinate (IMP), disodium guanylate (GMP), and their blends (E635) (Reineccius, 2006). Although some of these compounds are chemically identical to their naturally occurring counterparts, they are classified as synthetic when produced by nonnatural extraction methods (Britannica Editors., 2025). From a functional standpoint, both natural and synthetic enhancers act on the same taste receptors; however, natural enhancers often provide additional sensory attributes due to their complex composition, whereas synthetic enhancers deliver targeted and predictable flavor effects (Kurihara, 2009).

7.19.2 Safety aspects

From a toxicological and regulatory perspective, the safety of flavor enhancers is determined by chemical identity, purity, dosage, and exposure level, rather than by whether they are natural or synthetic (Reineccius, 2006). Regulatory authorities such as the FAO/WHO Joint Expert Committee on Food Additives (JECFA) and national food safety agencies evaluate flavor enhancers based on scientific risk assessments and establish acceptable daily intake (ADI) levels. Synthetic flavor enhancers such as MSG, IMP, and GMP have been extensively studied and are classified as safe for consumption when used within regulatory limits (Kurihara, 2009). Their defined chemical composition allows precise toxicological evaluation and consistent quality control (Vasilaki et al., 2021). Natural flavor enhancers, while often perceived as safer, are complex mixtures whose composition may vary depending on raw material quality, fermentation conditions, and processing methods. This variability can introduce challenges related to standardization, allergenicity, and batch-to-batch consistency (Fadimatou, Bonheur and Loh, 2024). Therefore, natural origin does not automatically imply superior safety, just as synthetic origin does not imply risk when proper regulatory standards are followed (Britannica Editors., 2025).

7.19.3 Consumer perception

Consumer perception represents one of the most significant differences between natural and synthetic flavor enhancers. Numerous sensory and consumer studies indicate that modern consumers strongly associate the terms *"natural," "clean-label,"* and *"fermented"* with health, safety, and authenticity (Prescott, 2015). Natural flavor enhancers are generally preferred by consumers seeking minimally processed foods and transparent ingredient labels. Ingredients such as yeast extract, mushroom powder, and seaweed extracts are often perceived as wholesome and acceptable, even though they perform the same function as synthetic enhancers (Vasilaki et al., 2021).

In contrast, synthetic flavor enhancers—particularly MSG—have faced negative consumer perceptions due to misinformation and historical controversy, despite strong scientific evidence supporting their safety (Kurihara, 2009). As a result, food manufacturers increasingly replace synthetic enhancers with natural alternatives in response to consumer demand, rather than scientific necessity (Prescott, 2015). However, sensory studies demonstrate that consumers often prefer products formulated with synthetic enhancers in applications where flavor consistency, intensity, and shelf stability are critical, such as snacks, instant foods, and beverages (Reineccius, 2006). This highlights a disconnect between perceived naturalness and actual sensory performance.

7.20 Nutraceutical benefits of flavor enhancers

Recent research indicates that flavor enhancers—particularly those derived from natural and fermentation-based sources—provide benefits beyond taste enhancement by contributing bioactive compounds with measurable nutraceutical properties. Studies on umami- and kokumi-active ingredients derived from mushrooms, yeast, fermented plant materials, and protein hydrolysates demonstrate antioxidant, antimicrobial, and metabolic-supporting effects linked to their amino acids, peptides, polyphenols, and Maillard reaction products (Amaliah et al., 2024; Vasilaki et al., 2021).

7.20.1 Antioxidant activity

One of the most consistently reported nutraceutical benefits of natural flavor enhancers is their antioxidant capacity. Fermentation-derived enhancers and protein hydrolysates contain bioactive peptides, polyphenols, and Maillard reaction products that scavenge free radicals and inhibit lipid oxidation. Hydrolysates derived from soybean, sunflower, and corn proteins have demonstrated significant antioxidant activity while simultaneously enhancing umami and mouthfeel (Habinshuti et al., 2019).

Similarly, natural umami ingredients extracted from mushrooms and plant leaves—such as shiitake by-products and mekai (*Albertisia spp.*) leaves—contain polyphenols and flavonoids that contribute to antioxidant protection in food systems (Harada-Padermo et al., 2020). These properties support oxidative stability in foods and may indirectly reduce dietary oxidative stress when such ingredients are consumed regularly.

7.20.2 Antiinflammatory potential

Although flavor enhancers are not medicinal agents, several natural umami sources contain compounds associated with reduced inflammatory responses. Mushroom-derived umami ingredients and fermented plant extracts contain polysaccharides, peptides, and secondary metabolites known to modulate oxidative and inflammatory pathways (Beluhan and Ranogajec, 2011; Vasilaki et al., 2021). Kokumi-active peptides identified in fermented dairy and legume products have been linked to improved sensory richness alongside modulation of cellular stress responses (Feng et al., 2016). The review on natural umami compounds emphasizes that these bioactive components often coexist with flavor-enhancing molecules, allowing foods to deliver both sensory and physiological value without additional additives (Amaliah et al., 2024).

7.20.3 Antidiabetic and metabolic-related support

Direct antidiabetic claims are not established for flavor enhancers; however, indirect metabolic benefits are documented. Natural umami enhancers enable significant sodium and fat reduction while maintaining palatability, which supports dietary strategies for managing hypertension and metabolic disorders (Vasilaki et al., 2021). In addition, fermentation-derived flavor enhancers contain peptides and amino acids that slow carbohydrate digestion and improve postmeal glycemic response by enhancing food structure and protein availability (Mottram, 2021). These effects contribute to better metabolic control when flavor enhancers are used to improve acceptance of balanced, low-salt diets.

7.20.4 Gut health and laxative-supporting effects

Several natural flavor enhancers originate from fermented foods and plant materials rich in dietary fiber, peptides, and organic acids. Fermented soy products, yeast extracts, and vegetable hydrolysates stimulate digestive secretions and support gut motility by activating taste receptors in the gastrointestinal tract (Kurihara, Kashiwayanagi, 2000).

While flavor enhancers are not laxatives, the improved digestion, increased saliva and gastric juice secretion, and enhanced intake of fiber-rich foods indirectly support bowel regularity. Plant-based umami ingredients such as mekai leaves and chaya leaves also contribute mild laxative-supporting effects due to their polyphenol and fiber content (Hutasingh et al., 2023).

7.20.5 Antimicrobial properties

Antimicrobial activity has been reported for several natural flavor-enhancing systems, particularly Maillard reaction products and fermented protein hydrolysates. These compounds inhibit the growth of spoilage and pathogenic microorganisms while enhancing flavor intensity (Habinshuti et al., 2019). Extracts from traditional umami-rich plants such as *Albertisia spp.* have demonstrated antimicrobial activity alongside their taste-enhancing function, highlighting their dual role as flavor and functional ingredients.

The future of natural flavors and natural flavor enhancers is closely aligned with the evolving expectations of consumers, regulatory bodies, and the food industry toward safer, cleaner, and more functional food systems. As global demand shifts toward minimally processed and label-friendly products, natural flavoring substances derived from plants, fermentation, and food by-products are expected to play an increasingly central role in food formulation. Scientific evidence suggests that natural umami- and kokumi-rich ingredients not only enhance sensory quality but also contribute bioactive compounds that support nutritional value and product stability (Amaliah et al., 2024; Vasilaki et al., 2021).

7.21 Conclusion

One of the most promising future directions lies in the integration of natural flavor enhancers with public health nutrition goals, particularly sodium and sugar reduction. Research has consistently shown that natural umami compounds can compensate for reduced salt levels while maintaining consumer acceptance, offering an effective strategy for addressing diet-related noncommunicable diseases (Vasilaki et al., 2021). As governments and health agencies intensify reformulation targets, natural flavor enhancers are likely to become key tools in developing healthier food products without compromising palatability. Advances in fermentation technology and bioprocessing are also expanding the scope of natural flavor development. Controlled fermentation and enzymatic hydrolysis enable the generation of flavor-rich peptides, amino acids, and nucleotides from plant proteins, algae, and agroindustrial by-products. This not only enhances flavor complexity but also supports sustainability by valorizing food waste streams (Habinshuti et al., 2019; Harada-Padermo et al., 2020). Such approaches align strongly with circular economy principles and are expected to gain wider industrial adoption.

From a nutraceutical perspective, the future research focus is likely to shift toward characterizing multifunctional flavor ingredients those that combine sensory enhancement with antioxidant, antimicrobial, and gut-supportive properties. Although flavor enhancers are not therapeutic agents, accumulating evidence indicates that natural flavor systems can contribute to improved oxidative stability, digestive stimulation, and dietary compliance when incorporated into balanced diets (Mottram, 2021; Amaliah et al., 2024). Further clinical and mechanistic studies will be essential to substantiate these health-supportive roles and establish safe, evidence-based applications. Consumer perception will continue to shape the trajectory of natural flavors. Growing skepticism toward synthetic additives has increased acceptance of ingredients such as yeast extracts, fermented plant products, seaweed, mushrooms, and protein hydrolysates. Transparent labeling and clear communication of ingredient origin will be critical in building consumer trust and ensuring regulatory compliance. In conclusion, natural flavors and natural flavor enhancers represent a convergence of sensory science, nutrition, and sustainability. Their future scope extends beyond taste enhancement toward supporting healthier diets, cleaner labels, and environmentally responsible food production. With continued interdisciplinary research and responsible innovation, natural flavor systems are well positioned to become foundational components of next-generation food products.

AI disclosure

During the preparation of this work, the authors used Open AI for grammar correction and paraphrasing and to check spelling errors. After using this tool/service, the authors reviewed and edited the content as needed and take full responsibility for the content of the publication.

References

Aggarwal, B.B., Harikumar, K.B., 2009. Potential therapeutic effects of curcumin, the anti-inflammatory agent, against neurodegenerative, cardiovascular, pulmonary, metabolic, autoimmune and neoplastic diseases. International Journal of Biochemistry and Cell Biology 41 (1), 40–59. Available at https://doi.org/10.1016/j.biocel.2008.06.010.

Amaliah, N., Mahendradatta, M., Zainal, Z., & Salengke, S. (2024) Trends in natural flavor enhancer: A review on umami compounds. In *BIO Web of Conferences*. EDP Sciences, Indonesia https://www.bio-conferences.org/ 96.

Asioli, D., Aschemann-Witzel, J., Caputo, V., Vecchio, R., Annunziata, A., Næs, T., et al., 2017. Making sense of the "clean label" trends: A review of consumer food choice behavior and discussion of industry implications. Food Research International 99, 58–71.

Belitz, H.D., Grosch, W., Schieberle, P., 2009. Food chemistry. Springer Berlin Heidelberg, Germany, pp. 1–1070. Available at https://doi.org/10.1007/978-3-540-69934-7, http://springerlink.com/openurl.asp?genre=book&isbn=978-3-540-69933-0.

Beluhan, S., Ranogajec, A., 2011. Chemical composition and non-volatile components of Croatian wild edible mushrooms. Food chemistry 124 (3), 1076–1082. Available at https://doi.org/10.1016/j.foodchem.2010.07.081.

Boland, A.B., Delahunty, C.M., van Ruth, S.M., 2013. Influence of food matrix on flavour release and perception. Food chemistry 141 (2), 1307–1315.

Britannica Editors. (2025). Flavouring: Natural and artificial flavourings and extracts. Encyclopaedia Britannica.

Burt, S., 2004. Essential oils and antibacterial properties. International Journal of Food Microbiology 94 (3), 223–253.

Chen, Y., Wang, X., Liu, S., 2024. Chemical and sensory quality of citrus-flavoured vegetable oils. Discover Food 4, 143–155.

Czerny, M., Wagner, R., Grosch, W., 2023. Structure–odor relationships of flavour compounds in plant foods. Journal of Agricultural and Food Chemistry 71 (18), 6894–6908.

Fadimatou, B., Bonheur, B., Loh, A.M.B., 2024. Sensory and nutritional properties and stability of formulated organic food flavour enhancers. Open Journal of Applied Sciences 14 (06), 1490–1506. Available at https://doi.org/10.4236/ojapps.2024.146099.

Feng, T., Zhang, Z., Zhuang, H., Zhou, J., Xu, Z., 2016. Effect of peptides on kokumi taste sensation. Mini-Reviews in Organic Chemistry 13 (4), 255–261.

Fernandes, B., Oliveira, M.C., Marques, A.C., dos Santos, R.G., Serrano, C., 2024. Microencapsulation of essential oils and oleoresins: Applications in food products. Foods 13 (23). Available at https://doi.org/10.3390/foods13233873, www.mdpi.com/journal/foods.

Food Safety Institute. (2023). Flavouring agents: Natural, nature-identical and artificial. Food Safety Institute Publications.

Habinshuti, I., Chen, X., Yu, J., et al., 2019. Antioxidant, antimicrobial and sensory properties of Maillard reaction products derived from protein hydrolysates. LWT – Food Science and Technology 101, 694–702.

Harada-Padermo, S.S., Dias-Faceto, L.S., Selani, M.M., et al., 2020. Umami ingredient produced from shiitake mushroom by-products. Food Research International 137, 109540.

Hossain, M.S., Wazed, M.A., Asha, S., et al., 2025. Flavor and well-being: A comprehensive review of food choices, nutrition, and health interactions. Fooad Science & Nutrition 13 (5), e70276.

Hubbs, A.F., Goldsmith, W.T., Kashon, M.L., Frazer, D., Mercer, R.R., Battelli, L.A., et al., 2008. Respiratory toxicologic pathology of inhaled diacetyl in Sprague-Dawley rats. Toxicologic Pathology 36 (2), 330–344. Available at https://doi.org/10.1177/0192623307312694.

Hutasingh, N., Chuntakaruk, H., Tubtimrattana, A., Ketngamkum, Y., Pewlong, P., Phaonakrop, N., et al., 2023. Metabolite profiling and identification of novel umami compounds in the chaya leaves of two species using multiplatform metabolomics. Food Chemistry 404 (Part A), 134564 Article.

Kim, S.A., Smith, S., Beauchamp, C., Song, Y., Chiang, M., et al., 2018. Cariogenic potential of sweet flavors in electronic-cigarette liquids. PLoS One 13 (9), e0203717. Available at https://doi.org/10.1371/journal.pone.0203717.

Kurihara, K. (2009).Glutamate: From discovery as a food flavor to role as a basic taste (umami). *American Journal of Clinical Nutrition*, 90, 719S–722S. http://www.ajcn.org/cgi/reprint/90/3/719S 90.

Kurihara, K., Kashiwayanagi, M., 2000. Basic characteristics of glutamate and umami sensing in the oral cavity and gut. Journal of Nutrition 130, 931S–934S.

Loussouarn, M., Krieger-Liszkay, A., Svilar, L., Bily, A., Birtić, S., Havaux, M., 2017. Carnosic acid and carnosol, two major antioxidants of rosemary, act through different mechanisms. Plant Physiology 175 (3), 1381–1394.

Lake, B.G., 1999. Coumarin metabolism and hepatotoxicity. Food and Chemical Toxicology 37 (4), 423–453.

Maga, J.A., Yamaguchi, S., 1983. Flavor enhancers. CRC Press.

Ma, F., Li, Y., Zhang, Y., Zhang, Q., Li, X., Cao, Q., et al., 2024. Effects of umami substances as taste enhancers on salt reduction in meat products: A review. Food Research International 185, 114248. Available at https://doi.org/10.1016/j.foodres.2024.114248.

McKay, D.L., Blumberg, J.B., 2006. A review of the bioactivity and potential health benefits of chamomile tea (*Matricaria recutita* L.). Phytotherapy Research 20 (7), 519–530. Available at https://doi.org/10.1002/ptr.1900.

Mortensen, G., Sørensen, A.D., 2023. Precision flavour formulation and sensory consistency in processed foods. Trends in Food Science & Technology 132, 240–249.

Mottram, D.S., 2021. Flavour formation in foods: Understanding sensory and nutritional implications. Food chemistry 360, 130038.

Nelson, G., Chandrashekar, J., Hoon, M.A., Feng, L., Zhao, G., Ryba, N.J.P., et al., 2002. An amino-acid taste receptor. Nature 416 (6877), 199–202. Available at https://doi.org/10.1038/nature726.

Ogori, A.F., 2022. Natural flavours: Sources, classification and applications in foods. Food Science and Nutrition Review 4 (3), 112–125.

Ohsu, T., Amino, Y., Nagasaki, H., Yamanaka, T., Takeshita, S., Hatanaka, T., et al., 2010. Involvement of the calcium-sensing receptor in human taste perception. Journal of Biological Chemistry 285 (2), 1016–1022. Available at https://doi.org/10.1074/jbc.M109.029165.

Parker, J.K., Elmore, J.S., Methven, L., 2025. Perceptual interactions of aroma compounds and enhancement of flavour quality in foods. Trends in Food Science & Technology 165, 105315.

Perez-Jiménez, M., Chaya, C., Pozo-Bayón, M.Á., 2019. Individual differences and effect of phenolic compounds in the immediate and prolonged in-mouth aroma release and retronasal aroma intensity during wine tasting. Food Chemistry 285, 147–155.

Prescott, J., Young, O., 2024. Multisensory flavour perception and consumer preference. Current Opinion in Food Science 52, 101056.

Prescott, J., 2015. Flavour as a psychological construct: Implications for perception, preference and food choice. Flavour 4, 15.

Reshna, K.R., Gopi, S., Balakrishnan, P., 2022. Flavors and fragrances in food processing: Preparation and characterization methods. Flavors and fragrances in food processing. American Chemical Society, pp. 1–19. Available at https://doi.org/10.1021/bk-2022-1433.ch001.

Reineccius, G., 2006. Flavor chemistry and technology, 2nd ed. CRC Press.

Reshna, K.R., Gopi, S., Balakrishnan, P., 2023. Introduction to flavor and fragrance in food processing. ACS Symposium Series 1433, 1–19. Available at https://doi.org/10.1021/bk-2022-1433.ch001, https://pubs.acs.org/series/symposium.

Román, S., Sánchez-Siles, L.M., Siegrist, M., 2017. The importance of food naturalness for consumers. Trends in Food Science & Technology 67, 44–57.

De Santis, D., 2024. Food flavour chemistry and sensory evaluation. Foods 13 (5), 634.

Shahidi, F., Ambigaipalan, P., 2015. Phenolics and polyphenolics in foods, beverages and spices: Antioxidant activity and health effects—A review. Journal of Functional Foods 18, 820–897. Available at https://doi.org/10.1016/j.jff.2015.06.018.

Spence, C., 2023. Multisensory flavour perception and well-being. Food Quality and Preference 105, 104783.

Srinivasan, K., 2014. Antioxidant potential of spices and their active constituents. Critical Reviews in Food Science and Nutrition 54 (3), 352–372. Available at https://doi.org/10.1080/10408398.2011.585525.

Stone, H., Herbert, A., 2012. Sensory Evaluation Practices in Food Science. Elsevier Academic Press.

Vasilaki, A., Panagiotopoulou, E., Koupantsis, T., Katsanidis, E., Mourtzinos, I., 2021. Recent insights in flavor-enhancers. Critical Reviews in Food Science and Nutrition 61 (22), 3802–3817. Available at https://doi.org/10.1080/10408398.2021.1939264.

Walton, N.J., Mayer, M.J., & Narbad, A. 2003. Vanillin. Phytochemistry, 63 (5), 505–515. Available at https://doi.org/10.1016/s0031-9422(03)00149-3.

Yamaguchi, S., & Ninomiya, K. (2000). Umami and food palatability. *Journal of Nutrition, 4* http://jn.nutrition.org 130.

Yamaguchi, S., 1991. Basic properties of umami and its effects on food flavor. Food Reviews International 7 (3), 247–264.

Yamamoto, T., Inui-Yamamoto, C., 2023. The flavor-enhancing action of glutamate and its mechanism involving kokumi perception. Npj Science of Food 7, 3. Available at https://doi.org/10.1038/s41538-023-00178-2.

Zhang, F., Klebansky, B., Fine, R.M., Xu, H., Pronin, A., Liu, H., et al., 2008. Molecular mechanism for the umami taste synergism. Proceedings of the National Academy of Sciences 105 (52), 20930–20934. Available at https://doi.org/10.1073/pnas.0810174106.

Zhang, X., Wang, Q., Liu, Z., et al., 2023. Plant protein-based emulsifiers: Mechanisms, techniques for emulsification enhancement and applications. Food Hydrocolloids 144. Available at https://doi.org/10.1016/j.foodhyd.2023.109008 Article 109008.

Nutritional and nutraceutical colors

8

Shilpa P.[1], *B.K. Indu*[2], *and Sudipta Kumar Mohanty*[2]
[1]Department of Nutrition and Dietetics, Padmashree Institute of Management and Sciences, Bengaluru, Karnataka, India, [2]Department of Biotechnology, Padmashree Institute of Management and Sciences, Bengaluru, Karnataka, India

8.1 Introduction

Color is one of the most influential sensory attributes affecting how consumers perceive and accept food. Unlike taste or aroma, which requires direct interaction with the product, color is the first attribute observed and often determines initial purchase decisions. It provides visual cues about freshness, flavor, and quality. For instance, a bright red tomato is associated with ripeness and flavor, while browning in fruits often suggests spoilage or deterioration. Studies consistently show that color significantly influences consumer perception and acceptance, sometimes even more strongly than taste (Clydesdale, 1993; Downham and Collins, 2000). Beyond consumer perception, color is widely recognized as an important indicator of food quality. Food industries routinely consider color as a critical parameter in quality assessment because it reflects product composition, processing conditions, and storage stability. Visible colors changes often indicate chemical or biochemical reactions such as enzymatic browning, pigment degradation, or Maillard reactions (Pathare et al., 2013). For example, the fading of green color in leafy vegetables indicates chlorophyll breakdown, while browning in cut fruits reflects oxidative reactions. Such changes directly influence consumer acceptance and the market value of food products. From a processing standpoint, food colorants are often used to restore or maintain desirable hues that may be lost during manufacturing and storage. Exposure to heat, light, oxygen, or pH changes can degrade natural pigments and alter the visual appearance of foods. To maintain consistent product appearance, manufacturers frequently add colorants to standardize color across production batches (Silva et al., 2022). Color therefore functions not only as a natural quality indicator but also as a technological tool in food processing.

Colors also plays a crucial role in marketing and product identity. In competitive markets, visual appearance strongly influences purchasing decisions. Consumers often associate certain colors with specific flavors, for example, red with berry flavors, yellow with citrus, and green with mint or freshness. These associations help build product recognition and brand identity, particularly in beverages, confectionery, and snack products (Delgado-Vargas and Paredes-López, 2003). Growing awareness of food safety and health concerns has led to increasing interest in natural food colorants. Natural pigments such as carotenoids, anthocyanins, betalains, and chlorophylls are gaining attention not only for their coloring properties but also for their potential nutraceutical

Nutraceutical Applications in Food Processing and Preservation. DOI: https://doi.org/10.1016/B978-0-443-45784-5.00015-0

benefits, including antioxidant and antiinflammatory effects (Sigurdson et al., 2017). This dual functionality has elevated natural colorants from simple additives to functional ingredients in modern food systems. Colorants also help maintain consumer acceptance when foods are fortified with nutrients that may alter their appearance. For instance, beverages enriched with vitamins or minerals may appear visually different without color correction. By restoring expected color characteristics, colorants help maintain sensory appeal while supporting nutritional enhancement. Thus, color in foods operates at the intersection of consumer psychology, food quality assessment, industrial processing, and marketing. As awareness of health and sustainability increases, natural food colorants are increasingly recognized for their dual role as sensory enhancers and bioactive compounds in modern food systems.

8.2 Synthetic food colors

Synthetic food colors (also known as artificial or certified color additives) have been widely used for decades to provide consistent and intense hues in processed foods such as beverages, confectionery, bakery products, and sauces. These dyes are typically water-soluble compounds produced through chemical synthesis, many belonging to the azo dye family characterized by –N=N– bonds. Common examples include tartrazine (E102/FD&C Yellow No.5), sunset yellow FCF (E110/FD&C Yellow No.6), brilliant blue FCF (E133/FD&C Blue No.1), and Allura red AC (E129/FD&C Red No.40). These dyes offer high coloring strength, excellent stability during processing, and low production cost. Their use is regulated by food safety authorities such as the US Food and Drug Administration (FDA) and the European Food Safety Authority (EFSA), which establish acceptable daily intake levels and safety guidelines (Scotter, 2011; Silva et al., 2022).

8.2.1 Tartrazine (E102/FD&C Yellow No.5)

Tartrazine is one of the most widely used synthetic yellow azo dyes. It produces a lemon-yellow color and is commonly used in soft drinks, candies, snacks, sauces, and baked goods. The dye is valued for its strong coloring power and cost effectiveness. However, tartrazine has been associated with hypersensitivity reactions in susceptible individuals, including urticaria and asthma-like symptoms. The European Food Safety Authority (EFSA) reevaluated tartrazine in 2009 and established an acceptable daily intake (ADI) of 7.5 mg/kg body weight/day, concluding that it is safe within permitted limits. Nevertheless, labeling requirements remain in the European Union because of its association with the "Southampton Six" dyes linked to hyperactivity in children (McCann et al., 2007; EFSA, 2009a).

8.2.2 Sunset yellow FCF (E110/FD&C Yellow No.6)

Sunset yellow FCF is another commonly used azo dye that provides a bright orange color. It is widely used in beverages, confectionery, bakery products, and fruit preserves. The dye shows good stability under light and moderate heat conditions, making it suitable for beverages and syrups. Similar to tartrazine, sunset yellow has been examined for potential allergic reactions and behavioral effects in children. EFSA reassessed this dye in 2014 and established an ADI of 1 mg/kg body weight/day, concluding that it does not show genotoxic effects but may cause sensitivity in certain individuals (EFSA, 2014).

8.2.3 Brilliant blue FCF (E133/FD&C Blue No.1)

Brilliant blue FCF is a triphenylmethane dye that produces an intense blue color and is commonly used in beverages, dairy desserts, candies, and pharmaceutical products. It is often combined with other dyes to produce green or purple shades. Toxicological evaluations by JECFA and EFSA have concluded that Brilliant blue FCF is not genotoxic or carcinogenic at permitted intake levels, with an ADI of 6 mg/kg body weight/day (JECFA, 2010). Compared with several azo dyes, hypersensitivity reactions related to Brilliant blue are relatively rare (Scotter, 2011).

8.2.4 Allura red AC (E129/FD&C Red No.40)

Allura red AC is a widely used synthetic red azo dye that replaced the earlier banned dye Amaranth. It is commonly used in soft drinks, cereals, confectionery, and bakery products because of its stability to heat and light. Although considered safe within established ADI limits (7 mg/kg body weight/day), Allura red has also been included among dyes associated with behavioral concerns in children in certain studies (McCann et al., 2007; EFSA, 2009b).

8.3 Why industry still uses synthetic colors

Despite the increasing interest in natural pigments, synthetic dyes remain widely used in the food industry due to several technological advantages.

Cost and supply: Synthetic dyes can be produced at large scale with consistent purity, making them significantly less expensive than natural pigments.

Color intensity and consistency: These dyes provide strong hues even at very low concentrations, allowing manufacturers to achieve uniform color across production batches.

Processing stability: Synthetic colorants are generally more stable under conditions such as high temperature, light exposure, and pH variation. Many natural pigments degrade or change color under similar conditions.

For these reasons, synthetic dyes continue to be used in applications where cost efficiency, stability, and color uniformity are critical, even as demand for natural colorants grows.

8.4 Regulatory actions, labels, and bans on synthetic food colors

The use of synthetic food colors is tightly regulated worldwide due to ongoing concerns about their potential health risks, including allergenicity, behavioral effects in children, and carcinogenic findings in animal models. Regulatory frameworks vary between countries, reflecting differences in risk assessment approaches, consumer protection policies, and historical use. While most regions continue to permit a defined list of synthetic dyes within established Acceptable Daily Intakes (ADIs), there are significant differences in labeling requirements and bans. For example, the European Union mandates warning labels on certain azo dyes linked to hyperactivity (the "Southampton Six"), the United States has recently revoked the use of Red No. 3 (Erythrosine) in foods, and India continues to allow a broad range of synthetic colors but prohibits their use in infant foods. The Codex Alimentarius provides international benchmarks that many countries follow for trade harmonization.

8.5 Natural food colors

8.5.1 Definition and classification

Natural food colors are pigments obtained from biological sources such as plants, animals, algae, or microorganisms, used to impart or restore color in foods. These differ from synthetic colorants, which are chemically synthesized (Lourenço-Lopes et al., 2021). The most common classes of natural pigments include carotenoids, anthocyanins, betalains, bixin, and chlorophylls, each contributing distinct hues yellow/orange, red/purple/blue, red/violet, and green respectively (Carrillo et al., 2024). Many natural colorants have their applications in food industry (Table 8.1).

In contrast to artificial colorants, natural pigments are often accompanied by bioactive compounds, making them functionally beneficial beyond esthetics (Son, 2024). Some regulations also recognize "nature-identical" colors—synthetic versions of naturally occurring molecules—but many consumers and industries still prefer truly natural extracts (Lourenço-Lopes et al., 2021) (Tables 8.2 and 8.3).

8.5.2 Significance of natural colorants in the food industry

8.5.2.1 Sensory appeal and consumer perception

Color is a key determinant of food acceptability. It influences consumer expectations regarding flavor, freshness, and quality (EFSA, 2015). Even minor deviations from expected color may affect perceived taste or product freshness. Natural pigments are therefore used to replace or restore colors lost during processing, or to enhance naturally occurring hues (EFSA, 2025). With rising health awareness, consumers increasingly associate natural colors with purity and safety. The replacement of synthetic dyes by natural alternatives is now seen as a marker of premium quality and health-conscious manufacturing (Carrillo et al., 2024).

8.5.2.2 Clean-label movement and industry adaptation

The "clean-label" trend—focusing on recognizable, minimal, and safe ingredients—has significantly boosted demand for natural pigments. Studies show that consumers prefer front-of-pack claims like "colored with fruit and vegetable extracts" over "no artificial

Table 8.1 Health effects and evidence-based evaluation of common synthetic food colorants.

Color (name, code)	Common uses	Reported health concerns	Key references
Tartrazine (E102/ FD&C Yellow No.5)	Soft drinks, candies, snacks, sauces, baked goods	Hypersensitivity reactions; possible behavioral effects in children	EFSA (2009a); McCann et al. (2007)
Sunset Yellow FCF (E110/FD&C Yellow No.6)	Orange sodas, sweets, bakery, jams	Occasional allergic responses; possible hyperactivity in children	EFSA (2014); McCann et al. (2007)
Allura Red AC (E129/FD&C Red No.40)	Soft drinks, cereals, confectionery	Behavioral concerns; safe within ADI	EFSA (2009b); McCann et al. (2007)
Brilliant Blue FCF (E133/FD&C Blue No.1)	Beverages, candies, dairy desserts	Rare hypersensitivity reactions	JECFA (2010); Scotter (2011)
Ponceau 4 R (E124)	Jams, desserts, soft drinks	Hyperactivity concerns; banned in US	McCann et al. (2007)
Carmoisine (E122)	Confectionery, soft drinks	Allergic reactions; hyperactivity concerns	EFSA
Erythrosine (E127/ FD&C Red No.3)	Candied fruits, cake decorations	Thyroid tumor risk at high doses	FDA (2025)
Indigo Carmine (E132/FD&C Blue No.2)	Ice cream, candies	Tumor risk at very high doses in animals	JECFA (2010)
Quinoline Yellow (E104)	Soft drinks, confectionery	Hyperactivity concerns	McCann et al. (2007)

colors," suggesting a positive association with naturality rather than avoidance (Lourenço-Lopes et al., 2021). Food manufacturers face challenges in maintaining color stability because natural pigments are sensitive to light, heat, pH, and oxygen. However, emerging technologies like microencapsulation, nanoemulsion, and copigmentation are improving the stability and shelf life of natural pigments in beverages, confectionery, and dairy applications (Son, 2024; Carrillo et al., 2024). Additionally, sustainability and regional biodiversity are now being leveraged to source pigments such as curcumin from turmeric, betalains from beetroot, and anthocyanins from butterfly pea and purple corn (Lourenço-Lopes et al., 2021).

Natural pigments are increasingly recognized not only for their color but also for their nutraceutical or biofunctional properties. These compounds often act as antioxidants, antiinflammatory agents, and modulators of metabolic and cardiovascular health (Carrillo et al., 2024). Carotenoids such as β-carotene, lutein, and

Table 8.2 Comparison of global regulations and permissible limits for synthetic food colors.

Region/authority	Permitted synthetic colors	Key labeling requirements	Bans/restrictions	References
India (FSSAI)	Tartrazine (E102), Sunset Yellow (E110), Carmoisine (E122), Ponceau 4 R (E124), Brilliant Blue FCF (E133), Indigo Carmine (E132), Fast Green FCF (E143), Erythrosine (E127)	Declaration on food labels as *"Contains permitted synthetic colours"* with INS/E-number or common name. Infant foods prohibited from containing synthetic colors	Max limit: 100 ppm (singly/combined) in most foods; banned in infant foods	FSSAI, 2011 (Food Products Standards & Food Additives Regulations)
European Union (EFSA, EC 1333/2008)	Broad list including Tartrazine (E102), Sunset Yellow (E110), Allura Red (E129), Brilliant Blue (E133), Quinoline Yellow (E104), Carmoisine (E122), Ponceau 4 R (E124)	Mandatory warning for "Southampton Six" dyes: *"May have an adverse effect on activity and attention in children."* Must declare E-number on labels	Amaranth (E123) banned; strict ADI-based approvals; re-evaluations ongoing	EFSA, 2009a; EC Regulation 1333/2008
United States (FDA)	Certified colors: FD&C Yellow 5 (Tartrazine), Yellow 6 (Sunset Yellow), Red 40 (Allura Red), Blue 1 (Brilliant Blue), Blue 2 (Indigo Carmine), Green 3, Red 3 (historically)	Colors must be listed by FD&C name (e.g., *Yellow 5*) on labels. Tartrazine specifically requires declaration due to hypersensitivity concerns	2025: FDA revoked Red No. 3 (Erythrosine) due to thyroid tumor evidence. No mandatory hyperactivity warnings unlike EU	FDA, 2025; FDA Color Additive Status List
Codex Alimentarius (JECFA/FAO-WHO)	International GSFA permits several synthetic dyes including Tartrazine, Sunset Yellow, Allura Red, Brilliant Blue, Indigo Carmine, Ponceau 4 R, Carmoisine	Countries adopting Codex must declare permitted colors; labeling recommendations harmonized with JECFA specifications	Amaranth and some dyes excluded; ADIs set by JECFA guide national limits	Codex GSFA (General Standard for Food Additives); JECFA evaluations

Table 8.3 Natural food colorants, sources, and applications in the food industry.

Natural colorant	Major natural source	Typical color	Applications in food industry	Reference
β-Carotene (carotenoids)	Carrot (*Daucus carota*), palm oil, algae	Yellow–orange	Margarine, dairy products, beverages, bakery products	Sigurdson et al. (2017)
Lycopene (carotenoids)	Tomato (*Solanum lycopersicum*), watermelon	Red	Sauces, beverages, meat products, processed foods	Sigurdson et al. (2017)
Annatto (bixin/norbixin)	Seeds of *Bixa orellana*	Yellow–orange	Butter, cheese, margarine, bakery products, snacks	Downham and Collins (2000)
Curcumin (curcuminoids)	Turmeric (*Curcuma longa*) rhizome	Bright yellow	Curry powders, dairy products, confectionery, beverages	Delgado-Vargas and Paredes-López (2003)
Anthocyanins	Berries, red cabbage, grapes, purple sweet potato	Red–purple–blue	Fruit beverages, jams, confectionery, yogurt, desserts	Sigurdson et al. (2017)
Betalains (betanin)	Beetroot (*Beta vulgaris*)	Red–violet	Ice cream, dairy desserts, beverages, meat substitutes	Azeredo (2009)
Chlorophyll/chlorophyllin	Spinach, parsley, nettle, alfalfa	Green	Confectionery, beverages, dairy products, soups	Sigurdson et al. (2017)
Paprika extract (capsanthin)	Red pepper (*Capsicum annuum*)	Orange–red	Meat products, sausages, sauces, snack seasonings	Delgado-Vargas and Paredes-López (2003)
hycocyanin	Blue-green algae (*Spirulina platensis*)	Blue	Chewing gum, confectionery, beverages, frozen desserts	Joshi et al. (2003)
Crocin	Saffron (*Crocus sativus*)	Yellow–orange	Rice dishes, bakery products, beverages, functional foods	Delgado-Vargas and Paredes-López (2003)
Carminic acid (carmine)	Cochineal insect (*Dactylopius coccus*)	Red	Processed meat products, beverages, confectionery	Downham and Collins (2000)

lycopene are potent antioxidants. β-Carotene functions as provitamin A, supporting visual and immune health, while lutein and zeaxanthin are crucial for macular protection (Lourenço-Lopes et al., 2021). Lycopene has shown potential in reducing oxidative stress and improving endothelial function (Son, 2024). Anthocyanins (found in berries, grapes, purple corn) and betalains (from beetroot and amaranth) display strong radical-scavenging capacities and may help modulate lipid metabolism and inflammation (Carrillo et al., 2024). They are also linked to improved cardiovascular outcomes and cognitive protection through enhanced antioxidant status and nitric oxide signaling. Chlorophylls and chlorophyll derivatives have demonstrated photoprotective and antimutagenic properties in both cellular and in vivo studies, although more mechanistic data are needed to validate these effects in human nutrition (Son, 2024). Recent evidence also indicates that dietary pigments can modulate the gut microbiome, improve microbial diversity and reducing markers of inflammation. For instance, anthocyanin-rich diets have been associated with an increase in beneficial *Bifidobacterium* and *Lactobacillus* species, linking colorants to gut-mediated health benefits (Carrillo et al., 2024). Thus, natural food colors are increasingly being regarded as functional ingredients—bridging the gap between sensory science and preventive nutrition.

8.6 Nutraceutical potential of natural colors and pigments

8.6.1 Antioxidant activity and reduction of oxidative stress

Natural pigments possess potent antioxidant properties through diverse chemical and biological mechanisms, making them important components in nutraceutical science. Carotenoids such as β-carotene, lycopene, and lutein/zeaxanthin quench singlet oxygen and scavenge peroxyl radicals due to their extended polyene conjugation systems, effectively halting lipid peroxidation in membranes. β-Carotene also functions as a provitamin A, indirectly supporting antioxidant enzyme expression via retinoid signaling (Lourenço-Lopes et al., 2021). Lycopene demonstrates high singlet oxygen quenching efficiency, preventing oxidative modification of LDL particles and reducing markers of lipid oxidation in human intervention studies. Oxygenated carotenoids like lutein and zeaxanthin accumulate in the retina, filtering blue light and protecting photoreceptor cells from reactive oxygen species (ROS) damage (Carrillo et al., 2024).

Anthocyanins—the water-soluble pigments in berries and purple vegetables—donate hydrogen atoms to neutralize ROS, chelate transition metals, and activate the Nrf2 pathway to enhance endogenous antioxidant defenses (Khoo et al., 2017). Similarly, betalains, particularly betanin from beetroot, exhibit strong electron-transfer activity and reduce oxidative stress biomarkers such as malondialdehyde (MDA) in animal and human models (Azeredo, 2009). Curcumin, the yellow pigment of turmeric, scavenges ROS while inducing antioxidant genes like HO-1 and SOD via Nrf2 activation, although its poor bioavailability requires advanced

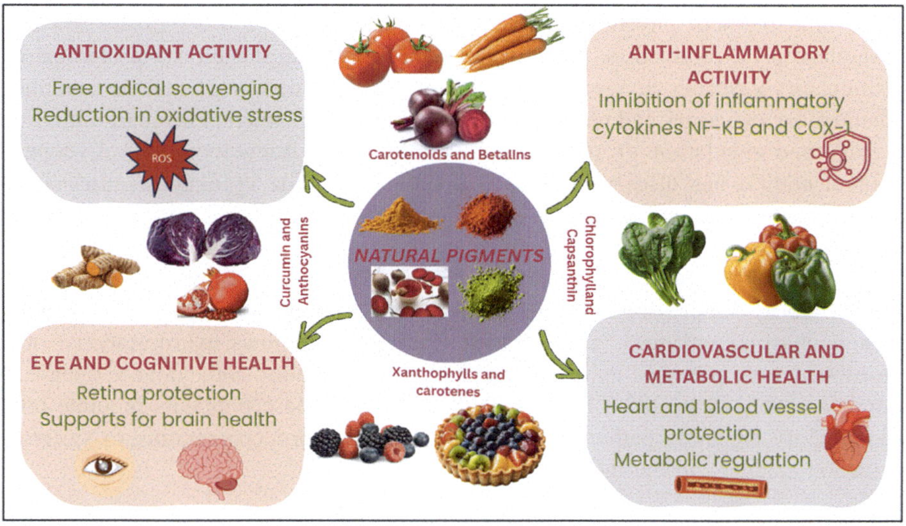

Figure 8.1 Nutraceutical benefits of natural food pigments. Major natural food pigments (carotenoids, anthocyanins, betalains, and chlorophyll) and their key health benefits including antioxidant, antiinflammatory, eye, cognitive, and cardiovascular protection.

delivery systems such as nanoformulations or phospholipid complexes (Hewlings and Kalman, 2017).

Astaxanthin, a marine carotenoid, spans cell membranes due to its polar ionone rings, enabling efficient lipid radical quenching and protection of mitochondrial membranes (Takahashi et al., 2023). Phycocyanin from *Spirulina* also demonstrates strong radical-scavenging capacity through its bilin chromophore, reducing oxidative stress both in vitro and in vivo (Romay et al., 2021). Collectively, these pigments represent a broad spectrum of natural antioxidants capable of mitigating oxidative damage implicated in chronic disease and aging (Fig. 8.1).

8.6.2 Antiinflammatory effects

Natural colorants exhibit antiinflammatory properties through modulation of key signaling pathways involved in immune regulation and cytokine synthesis. Curcumin, one of the most extensively studied nutraceutical pigments, inhibits the transcription factor NF-κB, downregulates pro-inflammatory enzymes such as COX-2, and suppresses cytokines including TNF-α and IL-6, producing broad antiinflammatory effects demonstrated in preclinical and clinical studies (Hewlings and Kalman, 2017).

Anthocyanins exert similar effects by blocking MAPK and NF-κB pathways in vascular and immune cells, reducing circulating levels of C-reactive protein (CRP) and adhesion molecules, thereby protecting against endothelial inflammation (Khoo et al., 2017; Carrillo et al., 2024). Astaxanthin downregulates pro-inflammatory eicosanoid synthesis and macrophage activation, reducing oxidative-induced inflammation in animal models (Takahashi et al., 2023).

The blue pigment phycocyanin from *Spirulina* inhibits COX-2 expression and prostaglandin E2 production, blocking NF-κB nuclear translocation and attenuating inflammatory cascades in vivo (Romay et al., 2021). Likewise, betalains modulate inflammatory cytokines and reduce exercise-induced inflammation through inhibition of ROS and regulation of redox-sensitive pathways (Azeredo, 2009). Lycopene supplementation has also been shown to reduce vascular cytokine synthesis and systemic inflammation, improving endothelial function (Lourenço-Lopes et al., 2021).

Other natural pigments such as chlorophyll derivatives exhibit mild antiinflammatory effects by suppressing phase I enzyme activity and promoting detoxification through phase II enzymes, while crocin from saffron shows neuroprotective and antiinflammatory potential by inhibiting microglial activation and cytokine release in neuronal tissues (Son, 2024; EFSA, 2015). Together, these pigments regulate both inflammatory signaling pathways and oxidative mediators, contributing to systemic protection against chronic inflammation.

8.6.3 Eye health and cognitive support

Several natural pigments play important roles in maintaining visual and cognitive health through antioxidant and neuroprotective mechanisms. Lutein and zeaxanthin are the principal macular pigments in the retina, where they absorb harmful blue light and neutralize ROS, protecting photoreceptor membranes rich in polyunsaturated lipids from oxidative damage. Clinical supplementation increases macular pigment optical density and improves visual acuity and contrast sensitivity (Lourenço-Lopes et al., 2021).

β-Carotene, as a provitamin A source, supports the synthesis of retinal, essential for phototransduction and maintenance of epithelial integrity, thereby preventing night blindness and xerophthalmia. Anthocyanins, particularly from bilberry and blackcurrant, enhance retinal microcirculation, improve rhodopsin regeneration, and reduce oxidative stress in photoreceptors, which contributes to reduced visual fatigue (Khoo et al., 2017; Carrillo et al., 2024).

Marine-derived astaxanthin can cross both the blood–retina and blood–brain barriers, stabilizing neuronal membranes, preventing mitochondrial oxidative injury, and supporting synaptic function (Takahashi et al., 2023). Curcumin also acts as a neuroprotective compound by inhibiting amyloid-β aggregation and microglial activation while enhancing synaptic plasticity (Hewlings and Kalman, 2017). Phycocyanin reduces glutamate-induced neurotoxicity and improves neuronal survival through antioxidant and antiapoptotic mechanisms (Romay et al., 2021; Bathaie and Farajzade, 2012).

Crocin from saffron exhibits mood-enhancing and cognition-supporting effects through modulation of serotonin and dopamine pathways along with protection against oxidative neuronal damage (Bathaie and Farajzade, 2012). Zeaxanthin esters from saffron and wolfberry further contribute to macular pigment density and visual protection, showing potential in preventing age-related macular degeneration (Lourenço-Lopes et al., 2021).

8.6.4 *Chronic disease prevention: cardiovascular and metabolic health*

Natural pigments contribute to the prevention of cardiovascular and metabolic disorders through mechanisms involving oxidative stress reduction, lipid modulation, and metabolic regulation. Anthocyanins improve endothelial nitric oxide bioavailability, reduce blood pressure, enhance vascular elasticity, and lower LDL oxidation, thereby improving lipid profiles and vascular health (Khoo et al., 2017; Carrillo et al., 2024).

Lycopene intake has been associated with reduced atherosclerotic progression and improved vascular reactivity due to its antioxidative and antiinflammatory effects on vascular tissues (Lourenço-Lopes et al., 2021). Marine carotenoids such as fucoxanthin and astaxanthin exhibit antiobesity and antidiabetic effects by enhancing mitochondrial fatty acid oxidation, upregulating uncoupling protein 1 (UCP1) in adipose tissue, and improving insulin sensitivity (Takahashi et al., 2023).

Curcumin modulates metabolic pathways by activating AMP-activated protein kinase (AMPK) and inhibiting NF-κB, improving glucose uptake and reducing inflammatory adipokines in metabolic syndrome models (Hewlings and Kalman, 2017). Phycocyanin and other *Spirulina* components improve lipid metabolism by reducing LDL cholesterol and triglycerides while enhancing antioxidant status and lowering blood pressure in small human trials (Romay et al., 2021).

Beetroot betalains, in combination with dietary nitrates, improve endothelial nitric oxide availability, reduce blood pressure, and enhance exercise endurance through vascular relaxation and antioxidant activity (Azeredo, 2009; Carrillo et al., 2024). Theaflavins from black tea also contribute to cardiovascular protection by inhibiting LDL oxidation, improving endothelial function, and enhancing insulin sensitivity (Khan and Mukhtar, 2019). These findings demonstrate that natural pigments contribute to both visual appeal and measurable metabolic health benefits.

8.7 Challenges of natural food colors

Despite the growing consumer and regulatory preference for natural colorants, several scientific and technological challenges limit their widespread replacement of synthetic dyes. One major limitation is their instability under processing and storage conditions. Pigments such as anthocyanins, betalains, and chlorophylls are sensitive to environmental factors including pH, temperature, light, and oxygen (Lourenço-Lopes et al., 2021). Anthocyanins, for example, display color variation depending on pH, complicating their application in multicomponent foods such as yogurts, beverages, and confectionery (Khoo et al., 2017). Betalains degrade under heat and light exposure, while chlorophylls undergo pheophytinization in acidic environments, resulting in dull olive-brown colors (Son, 2024). Technological approaches such as microencapsulation, copigmentation, nanoemulsions, and antioxidant stabilization have been explored to improve pigment stability during processing and storage (Carrillo et al., 2024). However, these methods often increase production costs and require specialized processing conditions.

References

Azeredo, H.M.C., 2009. Betalains: Properties, sources, applications, and stability—A review. International Journal of Food Science and Technology 44 (12), 2365–2376. Available at https://doi.org/10.1111/j.1365-2621.2007.01668.x, http://onlinelibrary.wiley.com/journal/10.1111/(ISSN)1365-2621.

Bathaie, S.Z., Farajzade, A., 2012. Saffron and its major ingredients: Recent advances in neuropharmacological studies. Iranian Journal of Basic Medical Sciences 15 (6), 1073–1086.

Carrillo, L., Yang, L., Luan, H., Zhang, X., Zhao, L., 2024. Natural pigments in the food industry: Enhancing stability, nutritional benefits, and gut microbiome health. Food Chemistry 460. Available at https://doi.org/10.1016/j.foodchem.2024.140514, Article 140514.

Clydesdale, F.M., 1993. Color as a factor in food choice. Critical Reviews in Food Science and Nutrition 33 (1), 83–101. Available at https://doi.org/10.1080/10408399309527614.

Delgado-Vargas, F., Paredes-López, O., 2003. Natural colorants for food and nutraceutical uses. CRC Press.

Delgado-Vargas, F., Luevanos, C., Gonzalez, C.A., 2024. Microbial biotechnology for the sustainable production of natural pigments: Current advances and future prospects. Trends in Food Science & Technology 152, 104233. Available at https://doi.org/10.1016/j.tifs.2024.104233.

Downham, A., Collins, P., 2000. Colouring our foods in the last and next millennium. International Journal of Food Science & Technology 35 (1), 5–22. Available at https://doi.org/10.1046/j.1365-2621.2000.00373.x.

European Food Safety Authority (EFSA), 2009a. Scientific opinion on the re-evaluation of Tartrazine (E 102). EFSA Journal 7 (11), 1331. Available at https://doi.org/10.2903/j.efsa.2009.1331.

European Food Safety Authority (EFSA), 2009b. Scientific opinion on the re-evaluation of Allura Red AC (E 129). EFSA Journal 7 (11), 1327. Available at https://doi.org/10.2903/j.efsa.2009.1327.

European Food Safety Authority (EFSA), 2014. Scientific opinion on the re-evaluation of Sunset Yellow FCF (E 110). EFSA Journal 12 (6), 3765. Available at https://doi.org/10.2903/j.efsa.2014.3765.

European Food Safety Authority (EFSA), 2015. Re-evaluation of chlorophylls (E 140(i)) and chlorophyllins (E 140(ii)) as food additives. EFSA Journal 13 (6), 4085. Available at https://doi.org/10.2903/j.efsa.2015.4085.

European Food Safety Authority (EFSA), 2023. Guidance on the safety evaluation of food additives produced by fermentation. EFSA Journal 21 (2), e07892. Available at https://doi.org/10.2903/j.efsa.2023.7892.

European Food Safety Authority (EFSA). 2025. Food colours. Retrieved from https://www.efsa.europa.eu/en/topics/topic/food-colours.

Hewlings, S.J., Kalman, D.S., 2017. Curcumin: A review of its effects on human health. Foods 6 (10). Available at https://doi.org/10.3390/foods6100092, https://www.mdpi.com/2304-8158/6/10/92/pdf.

JECFA. (2010). Evaluation of certain food additives. WHO Technical Report Series No. 966.

Joshi, V.K., Attri, D., Bala, A., Bhushan, S., 2003. Microbial pigments. Indian Journal of Biotechnology 2 (3), 362–369.

Khan, N., Mukhtar, H., 2019. Tea polyphenols in promotion of human health. Nutrients 11 (1), 39. Available at https://doi.org/10.3390/nu11010039.

Khoo, H.E., Azlan, A., Tang, S.T., Lim, S.M., 2017. Anthocyanidins and anthocyanins: Colored pigments as food, pharmaceutical ingredients, and the potential health benefits. Food & Nutrition Research 61 (1), 1361779. Available at https://doi.org/10.1080/16546628.2017.1361779.

Lourenço-Lopes, C., Fraga-Corral, M., Jimenez-Lopez, C., Pereira, A.G., Garcia-Oliveira, P., Simal-Gandara, J., 2021. Natural food colorants from algae: Phycobiliproteins and carotenoids—Sources, extraction, and applications. Food Chemistry 358, 129891. Available at https://doi.org/10.1016/j.foodchem.2021.129891.

McCann, D., Barrett, A., Cooper, A., Crumpler, D., Dalen, L., Grimshaw, K., et al., 2007. Food additives and hyperactive behaviour in 3-year-old and 8/9-year-old children. The Lancet 370 (9598), 1560–1567. Available at https://doi.org/10.1016/S0140-6736(07)61306-3.

Mertz, D., Hossain, M.S., Lee, S.Y., 2023. Synthetic biology for sustainable production of natural pigments: From pathway design to industrial application. Biotechnology Advances 66, 108215. Available at https://doi.org/10.1016/j.biotechadv.2023.108215.

Özkan, G., Bilek, S.E., 2022. Encapsulation technologies for natural food colorants: A review. Food Reviews International 38 (8), 1569–1590. Available at https://doi.org/10.1080/87559129.2020.1791889.

Pathare, P.B., Opara, U.L., Al-Said, F.A., 2013. Colour measurement and analysis in fresh and processed foods: A review. Food and Bioprocess Technology 6 (1), 36–60. Available at https://doi.org/10.1007/s11947-012-0867-9.

Romay, C., González, R., Ledón, N., Remirez, D., Rimbau, V., 2021. C-Phycocyanin: A biliprotein with antioxidant, anti-inflammatory and neuroprotective effects. Current Protein & Peptide Science 22 (6), 540–554. Available at https://doi.org/10.2174/1389203722666210308152433.

Scotter, M.J., 2011. Methods for the determination of European Union-permitted added natural colours in foods: A review. Food Additives & Contaminants: Part A 28 (5), 527–596. Available at https://doi.org/10.1080/19440049.2011.555844.

Sigurdson, G.T., Tang, P., Giusti, M.M., 2017. Natural colorants: Food colorants from natural sources. Annual Review of Food Science and Technology 8, 261–280. Available at https://doi.org/10.1146/annurev-food-030216-025923.

Silva, M.M., Reboredo, F.H., Lidon, F.C., 2022. Food colour additives: A synoptical overview on their chemical properties, applications in food products, and health side effects. Foods 11 (3), 379. Available at https://doi.org/10.3390/foods11030379.

Son, E., 2024. Pharmacological properties and stability of natural–colored foods: A literature review. Nutrition & Food Science 54 (8), 1463–1477. Available at https://doi.org/10.1108/nfs-04-2024-0114.

Takahashi, K., Okada, T., Oshima, Y., 2023. Fucoxanthin and astaxanthin: Marine carotenoids as promising functional ingredients. Marine Drugs 21 (5), 295. Available at https://doi.org/10.3390/md21050295.

Natural gums and gelling agents

9

B.K. Indu[1], *Chandan Hebbale*[2,3], *and Sudipta Kumar Mohanty*[1]
[1]Department of Biotechnology, Padmashree Institute of Management and Sciences, Bengaluru, Karnataka, India, [2]Brown University, Providence, RI, United States, [3]Brookside Botanical Gardens, Takoma Park, MD, United States

9.1 Introduction

Texture, stability, and mouthfeel strongly influence perceived food quality and consumer acceptance. Many processed foods face stability challenges during storage, including phase separation, sedimentation, creaming, and water expulsion from gel structures. Natural gums and gelling agents—collectively termed hydrocolloids—provide practical solutions by modifying rheological behavior and microstructure even at low usage levels. Hydrocolloids contribute to viscosity development, suspension stabilization, emulsion control, and gel network formation, thereby improving process tolerance and shelf-life (Nishinari, 2009).

The contemporary relevance of natural gums is closely linked with current formulation goals: reduction of sugar and fat without compromising texture, clean-label ingredient selection, and development of functional foods requiring stable delivery of bioactive ingredients. Hydrocolloid science is therefore integral to modern food engineering and nutraceutical product development (Saha and Bhattacharya, 2010; Phillips and Williams, 2021).

9.1.1 Hydrocolloids: concept and key terminology

Hydrocolloids are water-compatible, high-molecular-mass biopolymers—predominantly polysaccharides and, in some cases, proteins—that hydrate in aqueous environments to produce viscous dispersions or gelled structures. In food processing, these materials are used primarily to tailor rheological behavior, with applications that include thickening, stabilization of multiphase systems, emulsification support, and gel network formation. In this chapter, gums denote hydrocolloids employed mainly to increase viscosity and enhance dispersion stability, whereas gelling agents refer to hydrocolloids capable of creating three-dimensional networks that immobilize water and generate semi-solid texture.

From a practical standpoint, viscosity reflects the resistance of a food to flow and influences both sensory perception and industrial handling (mixing, pumping, and filling). Syneresis describes the release of liquid from a gel matrix, frequently observed as whey separation in fermented dairy products. Several hydrocolloid solutions exhibit shear-thinning (pseudoplastic) flow, in which apparent viscosity decreases with increasing shear rate—an attribute that enables stability at rest while allowing convenient pouring and desirable mouthfeel during consumption.

Nutraceutical Applications in Food Processing and Preservation. DOI: https://doi.org/10.1016/B978-0-443-45784-5.00009-5

9.2 Classification of natural gums and gelling agents

Natural gums and gelling agents can be classified by biological origin and chemical nature. This classification is helpful for predicting functional behavior and selecting appropriate hydrocolloids for specific food applications.

9.2.1 Classification by biological origin

(a) Plant exudate gums
- Gum arabic (Acacia gum)

(b) Seed/endosperm gums
- Guar gum
- Locust bean gum (carob gum)

(c) Fruit and plant cell-wall polysaccharides
- Pectin

(d) Seaweed-derived hydrocolloids
- Alginate (brown algae)
- Carrageenan (red algae)

(e) Microbial fermentation gums
- Xanthan gum (*Xanthomonas campestris*)

(f) Animal-derived hydrocolloids
- Gelatin (collagen derivative)

9.2.2 Classification by ionic character

Ionic character governs hydration kinetics, salt sensitivity, and interactions with proteins.

- Neutral hydrocolloids: guar gum, locust bean gum
- Anionic hydrocolloids: pectin, alginate, carrageenan, xanthan gum

Electrostatic interactions become particularly important in protein-containing systems such as milk beverages, yogurt, and formulated nutritional drinks.

9.3 Molecular basis of functionality: structure–property relationships

Hydrocolloid behavior in foods is dictated by polymer chain architecture, including molecular weight, branching patterns, charge density, and substituent groups. These structural parameters control the extent of hydration and swelling, chain entanglement, and the ability to form junction zones that stabilize gels.

9.4 Thickening mechanism

Viscosity enhancement by hydrocolloids arises largely from rapid hydration and swelling of polymer chains, followed by the development of an entangled macromolecular network within the continuous phase. Because hydrated polymer coils occupy a large effective volume, even low concentrations can markedly restrict water mobility and increase resistance to flow. The magnitude and stability of thickening depend on molecular weight, degree of branching, temperature history, and the ionic environment, as these factors govern chain expansion, intermolecular association, and shear response (Nishinari, 2009; Phillips and Williams, 2021).

9.5 Gelation mechanisms

Gelation involves transformation of a polymer dispersion into a structured system through establishment of a continuous three-dimensional network. Network formation can be achieved through several distinct pathways. In high-methoxyl pectin, gelation is typically promoted by low pH and elevated soluble solids, where reduced water availability and diminished electrostatic repulsion favor polymer–polymer association. In alginate and low-methoxyl pectin, gels form predominantly through calcium-mediated ionic crosslinking between carboxyl groups, producing junction zones that strengthen the network. Carrageenan gelation is associated with conformational ordering (helix formation) and subsequent aggregation of helices, with specific salts (e.g., K^+ or Ca^{2+}) modulating gel strength and elasticity. In gelatin systems, cooling promotes partial reformation of ordered helical segments that serve as physical crosslinks, resulting in thermoreversible gels. These mechanisms underpin the wide range of textures achievable with natural gelling agents (Saha and Bhattacharya, 2010; Rinaudo, 2008).

9.6 Natural gums: sources, chemistry, and functional properties

Natural gums are principally used as thickening and stabilizing agents and are essential for viscosity design and stability control in food systems such as sauces, beverages, dairy products, and bakery fillings.

9.6.1 Gum arabic (Acacia gum)

9.6.1.1 Source and processing

Gum arabic is a dried plant exudate obtained from Acacia species, particularly *Acacia senegal* and *Acacia seyal*. For food use, it is purified through dissolution, filtration, and spray drying to meet quality specifications.

9.6.1.2 Chemical structure and composition

Gum arabic consists largely of arabinogalactan polysaccharides with a glycoprotein fraction. The protein-associated fraction provides surface activity, allowing gum arabic to act as an effective natural emulsifier.

9.6.1.3 Functional properties

Gum arabic is characterized by excellent water solubility and comparatively low viscosity even at high solids concentration, enabling use in concentrated beverage emulsions. It provides effective emulsification and stabilization of oil-in-water systems due to its glycoprotein fraction and additionally exhibits film-forming capability, supporting encapsulation of flavors and fat-soluble micronutrients in spray-dried powders.

9.6.1.4 Food applications

Gum arabic is widely used for beverage emulsions (flavor oil stabilization), confectionery glazing, and encapsulation of flavors and vitamins via spray drying.

9.6.2 Guar gum (E412)

9.6.2.1 Source and extraction

Guar gum is derived from guar seeds (*Cyamopsis tetragonoloba*). Industrial production involves dehusking, separation of endosperm, milling, and purification.

9.6.2.2 Chemical nature

Guar gum is a galactomannan consisting of a mannose backbone with galactose side chains. This architecture creates strong hydration and rapid viscosity development.

9.6.2.3 Functional properties

Guar gum is a high-efficiency thickener, commonly providing substantial viscosity at low inclusion levels (typically 0.2%–0.5%, depending on matrix). It hydrates rapidly in cold water and contributes to freeze–thaw stability by limiting water migration and structural damage during ice crystal formation. In dairy and gel-like systems, guar gum enhances water binding and reduces syneresis, thereby improving product consistency during storage.

9.6.2.4 Food applications

Guar gum is used in ice cream and frozen desserts to improve body and melting resistance, in sauces and gravies to control viscosity, and in bakery systems (including gluten-free formulations) to improve water retention and dough structure.

A comprehensive safety reevaluation by EFSA supports guar gum as safe for use as a food additive when meeting identity and purity specifications (Mortensen et al., 2017).

9.6.3 Locust bean gum (carob gum; E410)

9.6.3.1 Source and chemistry

Locust bean gum is obtained from seeds of the carob tree (*Ceratonia siliqua*). Like guar gum, it is a galactomannan but has fewer galactose side chains, which influences its hydration profile and interactive behavior.

9.6.3.2 Functional role and synergy

Locust bean gum hydrates more slowly than guar gum, but it exhibits exceptional synergy with several gelling systems. In dairy matrices, combinations of locust bean gum with carrageenan improve gel strength and reduce syneresis. In thickened sauces, blends with xanthan gum generate elastic, stable viscosity.

9.6.3.3 Food applications

Locust bean gum is extensively applied in dairy desserts, ice cream, processed cheese, and sauce systems requiring enhanced texture and stability.

9.6.4 Xanthan gum (E415)

9.6.4.1 Source and production

Xanthan gum is an extracellular polysaccharide produced industrially through aerobic fermentation using Xanthomonas campestris. Following fermentation, xanthan is recovered by precipitation (commonly with alcohol), washed, dried, and milled to obtain a free-flowing powder suitable for food use. The fermentation route offers consistent quality and functionality, contributing to xanthan's widespread industrial adoption in diverse food categories.

9.6.4.2 Molecular structure and functional significance

Xanthan gum possesses a cellulose-like β-(1→4)-D-glucose backbone with trisaccharide side chains containing mannose and glucuronic acid residues. The presence of carboxyl groups imparts an anionic character. This architecture promotes formation of ordered conformations and yields exceptional viscosity stability under challenging formulation conditions.

A distinctive characteristic of xanthan gum is its pseudoplastic (shear-thinning) rheology, where viscosity decreases as shear rate increases (Garcia-Ochoa et al., 2000; Phillips and Williams, 2021). Technologically, this provides high product thickness at rest (preventing separation) while allowing easy pouring, pumping, and mouthfeel during consumption.

9.6.4.3 Stability and performance in food matrices

Xanthan gum demonstrates exceptional functional stability across a wide range of formulation and processing conditions. It retains viscosity over broad pH ranges, including highly acidic food systems, and remains effective in the presence of salts and minerals where many other hydrocolloids show loss of functionality. Xanthan solutions also tolerate thermal processing and high shear operations, providing consistent performance during mixing, pumping, and filling. In addition, xanthan gum contributes to freeze–thaw stability by limiting water mobility and reducing phase instability during temperature cycling. These combined properties make xanthan gum particularly suitable for emulsions and suspensions requiring long-term physical stability under variable storage and distribution environments.

9.6.4.4 Food applications

Xanthan gum is widely applied in salad dressings and emulsified sauces to enhance stability and provide controlled pourability, and in ketchup-type condiments to achieve yield stress, suspension stability and product cling. It is also used in beverages to maintain uniform suspension of particles and clouding agents, and in gluten-free bakery systems where viscosity development improves dough structure, gas retention and final product texture.

9.7 Natural gelling agents: sources, chemistry, and gelation behavior

Gelling agents are hydrocolloids capable of forming three-dimensional networks that entrap water and produce semi-solid structures. Gel formation may be thermoreversible or irreversible and depends on polymer type, concentration, ionic environment, pH, and the presence of cosolutes such as sugars.

9.7.1 Pectin (E440)

9.7.1.1 Source and extraction

Pectin is a structural polysaccharide present in the middle lamella and primary cell walls of plants. Industrially, it is extracted mainly from citrus peel and apple pomace—by-products of juice processing—through hot acid extraction, followed by filtration, concentration, precipitation, and drying. The sustainable sourcing from agro-industrial residues has supported its strong global market (Voragen et al., 2009; Thakur et al., 1997).

9.7.1.2 Chemical structure and classification

Pectin is composed primarily of α-(1→4)-linked D-galacturonic acid units, partially esterified with methanol. The degree of esterification (DE) is the principal determinant of gelling behavior and is used to classify pectin into:

- High methoxyl (HM) pectin (DE $> 50\%$)
- Low methoxyl (LM) pectin (DE $< 50\%$)

9.7.1.3 Gelation mechanisms

HM pectin gelation: HM pectin forms gels in the presence of high soluble solids (typically sucrose) and acidic pH. In this system, sugar reduces water availability for pectin hydration, allowing polymer–polymer interactions, while acid reduces electrostatic repulsion by decreasing ionization of carboxyl groups.

LM pectin gelation: LM pectin forms gels via ionic crosslinking with divalent ions, especially Ca^{2+}, through the well-known "egg-box" model. This mechanism is particularly valuable in low-sugar and sugar-free gels, supporting formulation for diabetic-friendly or calorie-reduced products.

9.7.1.4 Food applications

Pectin is extensively utilized in fruit-based and acidic food systems due to its reliable gelation and thickening performance. High methoxyl pectin remains the preferred gelling agent for traditional jams, jellies, marmalades, and fruit preserves formulated with high soluble solids and acidic pH, where it produces firm, glossy gels with characteristic spreadability. Low methoxyl pectin is widely applied in reduced-sugar and sugar-free fruit gels and dessert formulations because calcium-mediated gelation enables structured products without requiring high sucrose levels. Beyond preserves, pectin is commonly incorporated into fruit preparations for yogurt and fermented dairy products, where it improves viscosity, stabilizes fruit dispersion, and reduces serum separation during storage. Pectin also supports improved texture in bakery fruit fillings and beverage-type fruit systems, while contributing to clean-label positioning and fiber-associated functional attributes (Voragen et al., 2009; Sriamornsak, 2007).

9.7.2 Alginate (E401–E405)

9.7.2.1 Source and industrial significance

Alginate is obtained from brown seaweeds (e.g., Laminaria, Macrocystis, Ascophyllum). Extraction typically involves alkaline solubilization, filtration, and precipitation as alginic acid or calcium alginate, followed by conversion into sodium alginate or related salts.

9.7.2.2 Structure and gelation behavior

Alginate consists of (1→4)-linked β-D-mannuronic acid (M) and α-L-guluronic acid (G) residues arranged in blocks. The ratio and distribution of M and G blocks influence gel strength:

- High G-block content → stronger, more brittle gels
- Higher M-block content → softer, more elastic gels

Alginate gels form rapidly in presence of Ca^{2+} through ionic crosslinking, producing networks without heat (Grant et al., 1973; Draget et al., 2005). This feature enables cold-set gels suitable for sensitive nutrients and bioactives.

9.7.2.3 Food applications

Alginate is widely used in food processing because it forms gels rapidly through calcium-mediated ionic crosslinking, often without the need for heating. This functionality enables production of cold-set gels for heat-sensitive ingredients and supports a wide range of structured applications. Alginate is frequently applied in restructured foods and fabricated textures, where controlled gelation improves shape retention and bite characteristics. It is also used in edible coatings and films due to its strong water-binding capacity and barrier properties, assisting in moisture control and extension of shelf-life in fresh-cut fruits and vegetables. In functional food development, alginate is a preferred encapsulation matrix for probiotics, enzymes, flavors, and micronutrients, offering protective gel structures that improve ingredient stability and targeted delivery. Additionally, alginate gelation is central to spherification-based applications, generating uniform gel beads or "pearls" with desirable sensory texture.

9.7.3 Carrageenan (E407 and E407a)

9.7.3.1 Source and composition

Carrageenans are sulfated galactans extracted from red seaweeds (Rhodophyceae). Commercial carrageenan includes different fractions based on sulfate content and gelation behavior.

9.7.3.2 Major carrageenan types and functionality

- κ (kappa) carrageenan: forms strong, brittle gels; enhanced by K^+ salts
- ι (iota) carrageenan: forms soft, elastic gels; enhanced by Ca^{2+} salts
- λ (lambda) carrageenan: nongelling thickener; provides viscosity

9.7.3.3 Interaction with dairy proteins

Carrageenan exhibits strong functional relevance in dairy systems due to interactions with casein micelles, contributing to:

- Suspension stability in chocolate milk
- Reduced phase separation
- Improved texture in dairy desserts

This interaction property is one reason carrageenan remains central to industrial dairy stabilization (Imeson, 2009; de Vries et al., 2002).

9.7.3.4 Food applications

Carrageenan is extensively applied in dairy and nondairy formulations due to its ability to modify viscosity, stabilize dispersed systems, and form gels with specific textural properties depending on the carrageenan fraction used. In flavored milk and chocolate milk, carrageenan is widely employed to maintain suspension stability by preventing sedimentation of cocoa particles and reducing phase separation through interaction with milk proteins. In dairy desserts such as puddings and custards, κ-carrageenan contributes to firm gel formation, whereas ι-carrageenan produces softer, more elastic textures, enabling tailored mouthfeel based on product requirements. Carrageenan is also used in plant-based dairy alternatives to improve stability and body, particularly in formulations prone to separation. Beyond dairy, carrageenan contributes to moisture retention and texture development in processed meat systems, where it supports water binding and improves sliceability and yield.

9.7.4 Gelatin (E441)

9.7.4.1 Source and preparation

Gelatin is a protein hydrocolloid derived from partial hydrolysis of collagen obtained from animal connective tissues (skin, bones, and tendons). Controlled processing yields gelatin with defined bloom strength, viscosity, and melting behavior.

9.7.4.2 Gelation characteristics and sensory significance

Gelatin forms thermoreversible gels: it dissolves upon heating and gels upon cooling. This property produces highly desirable sensory qualities, notably melt-in-mouth behavior, which is difficult to replicate with plant-based polysaccharide gels.

Gelatin gels involve formation of partial triple-helical junction zones during cooling, creating a flexible network entrapping water.

9.7.4.3 Food applications

Gelatin is extensively used in gummies, jellies, marshmallows and related confectionery products due to its thermoreversible gelation and characteristic melt-in-mouth texture. In dairy desserts such as mousse and panna cotta, gelatin contributes to smooth gel structure and desirable mouthfeel. It is also used as a clarifying agent in beverages such as wine and juices through fining applications, and as a functional ingredient in foams and encapsulation systems where it supports stabilization and controlled release of flavors and bioactives.

9.7.4.4 Limitations and label considerations

Gelatin is not suitable for vegetarian/vegan foods and may require halal/kosher certification depending on source and target market. As demand for plant-based alternatives increases, pectin and alginate systems are increasingly used to replace gelatin in fruit-gel and gummy categories.

9.7.5 Comparative evaluation of gums and gelling agents (selection logic)

Although both gums and gelling agents are hydrocolloids, their selection depends on target texture and food matrix behavior.

Gums (thickeners/stabilizers) such as guar and xanthan primarily control viscosity and stabilize emulsions/suspensions.

Gelling agents such as pectin, alginate, carrageenan, and gelatin provide structured gel networks.

In practice, combinations are commonly required because a single hydrocolloid may not deliver optimum rheology, stability, and sensory properties simultaneously.

9.8 Synergistic hydrocolloid systems and blend engineering

In industrial practice, hydrocolloids are rarely used as single ingredients; instead, binary or ternary gum systems are designed to achieve targeted rheology, stability, and sensory texture while ensuring cost-effectiveness. Synergy occurs when polymer chains with complementary structures interact to generate enhanced viscosity, gel strength, or water immobilization relative to individual hydrocolloids. Such interactions influence the formation of junction zones, polymer entanglement density, and microstructural stability, thereby reducing phase separation and syneresis during storage (Nishinari, 2009; Saha and Bhattacharya, 2010; Phillips and Williams, 2021).

9.8.1 Industrially important synergistic combinations

9.8.1.1 Xanthan gum–locust bean gum

The xanthan–LBG system is widely adopted due to strong synergistic viscosity enhancement and elastic texture development. Xanthan contributes high viscosity and shear-thinning behavior, while LBG strengthens the network through associative interactions that increase molecular connectivity. The blend is particularly useful in sauces, dressings, and structured emulsions requiring stability at rest but ease of flow during dispensing (Saha and Bhattacharya, 2010).

9.8.1.2 Carrageenan–locust bean gum

Carrageenan provides gel formation (κ and ι fractions), while LBG improves gel elasticity and reduces brittleness. In dairy desserts, this combination enhances water-binding, improves mouthfeel, and reduces syneresis, supporting consistent texture throughout shelf-life (Phillips and Williams, 2021).

9.8.1.3 Guar gum–xanthan gum

Guar gum hydrates rapidly and builds bulk viscosity, while xanthan provides stability under variations in pH, salts, and thermal processing. This blend is frequently used in gravies and soups requiring stable viscosity and resistance to separation during heating and cooling cycles (Saha and Bhattacharya, 2010).

9.8.1.4 Alginate–pectin hybrid systems

Alginate gels rapidly through calcium-mediated ionic crosslinking, while pectin contributes improved texture smoothness and water-holding. Hybrid gels are of increasing relevance in fruit-structured foods and encapsulation-based delivery systems (Phillips and Williams, 2021).

9.9 Applications in major food systems: matrix-specific guidance

The selection and performance of gums and gelling agents depend strongly on food matrix composition (protein, sugar, minerals, fat), processing (heat, shear), and storage environment. Consequently, matrix-specific formulation strategies are essential.

9.9.1 Dairy products and dairy-based beverages

9.9.1.1 Flavored milk and chocolate milk

Stability issues in chocolate milk include cocoa sedimentation and creaming. Carrageenan is extensively used because it interacts with milk proteins, improving particle suspension, reducing phase separation, and enhancing viscosity and mouthfeel (Phillips and Williams, 2021).

Common formulation strategies include:

- Carrageenan alone for suspension stability
- Carrageenan + LBG for improved gel-like mouthfeel and reduced syneresis
- Xanthan for acidic dairy beverages with stable viscosity

9.9.1.2 Yogurt and fermented dairy systems

Fermented dairy products may exhibit syneresis due to weak gel microstructure, particularly in low-fat yogurt. Pectin, guar gum, and their blends enhance water binding and stabilize the serum phase, thereby reducing whey separation. Pectin is widely preferred in fruit yogurts due to compatibility with acidic conditions and fruit solids (Saha and Bhattacharya, 2010).

9.9.1.3 Ice cream and frozen desserts

Hydrocolloids control texture and meltdown by limiting ice recrystallization and stabilizing the unfrozen phase. Guar gum contributes viscosity and body, LBG improves creaminess and heat shock resistance, and blends provide robust freeze–thaw stability (Phillips and Williams, 2021).

9.9.2 Bakery products and gluten-free formulations

In wheat bakery products, gluten provides viscoelasticity and gas retention. In gluten-free systems, hydrocolloids serve as structuring agents to increase batter viscosity, improve dough handling, reduce staling, and enhance crumb softness. Xanthan gum and guar gum remain dominant due to effective viscosity control and water retention. Proper optimization is required because excessive hydrocolloid concentration can reduce volume and yield dense texture (Saha and Bhattacharya, 2010) (Table 9.1).

Table 9.1 Major natural gums and gelling agents: origin, key chemistry, and principal food functions.

Hydrocolloid	Source	Chemical nature	Primary role	Typical applications
Gum arabic	Acacia exudate	Polysaccharide + protein fraction	Emulsifier, encapsulant	Beverage emulsions, spray-dried flavors
Guar gum (E412)	Guar seed endosperm	Galactomannan	Thickener	Ice cream, sauces, bakery
Locust bean gum (E410)	Carob seed	Galactomannan	Stabilizer; synergy	Dairy desserts, ice cream
Xanthan gum (E415)	Fermentation	Anionic polysaccharide	Shear-thinning thickener	Dressings, ketchup, gluten-free dough
Pectin (E440)	Citrus/apple	Polygalacturonic acid	Fruit gel; thickener	Jams, yogurt fruit base
alginate (E401–E405)	Brown seaweed	uronic acids	Ca^{2+} gel; encapsulation	Coatings, beads, spherification
Carrageenan (E407)	Red seaweed	Sulfated galactan	Dairy stabilizer; gel	Chocolate milk, puddings
Gelatin (E441)	Collagen	Protein	Thermoreversible gel	Gummies, desserts

9.9.3 Sauces, soups, gravies, and salad dressings

These products require stability at rest with controlled flow during dispensing. Xanthan gum provides shear-thinning behavior and yield stress, enabling suspension of particulates and stable viscosity across processing conditions. Guar gum is a cost-effective thickener but may be less stable under extreme thermal and shear conditions, hence blends (guar–xanthan) are commonly adopted (Saha and Bhattacharya, 2010) (Fig. 9.1.

9.9.4 Fruit gels, jams, jellies, and confectionery

Pectin remains the most important gelling agent for fruit products. HM pectin supports conventional high-sugar gelation at acidic pH, whereas LM pectin enables calcium-set gels appropriate for low-sugar and sugar-free products. In confectionery, gelatin provides melt-in-mouth gels; plant-based alternatives often use pectin or carrageenan depending on desired texture (Phillips and Williams, 2021).

9.9.5 Beverage emulsions and encapsulated ingredients

Gum arabic is the primary natural emulsifier for beverage flavor oils due to its protein-associated surface activity and exceptional solubility. It is also widely used for spray-dried encapsulation of flavors and fat-soluble vitamins. Xanthan gum is useful for particle suspension and cloud stability in beverages requiring long-term physical stability (Saha and Bhattacharya, 2010).

9.10 Nutritional and functional food relevance

Natural gums and gelling agents are not only technological additives but also contribute dietary fiber-like effects when used at adequate amounts. Viscosity is a key physiological factor influencing gastric emptying and postprandial glycemic response (Ellis et al., 1996). Hydrocolloids further support development of low-fat and low-sugar foods by restoring mouthfeel and improving sensory acceptance, thereby enabling reformulation strategies relevant to obesity prevention and metabolic health (Saha and Bhattacharya, 2010).

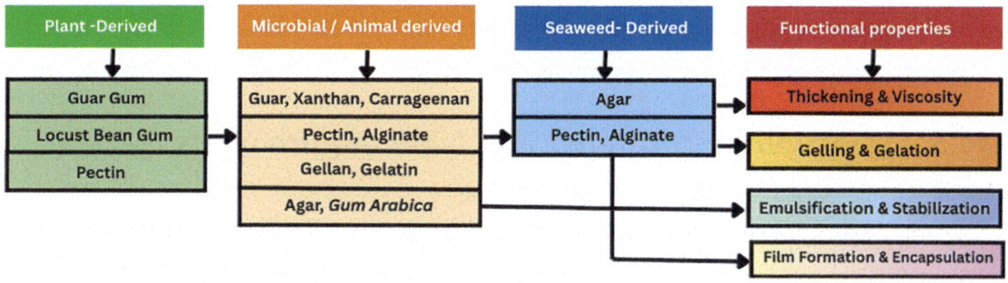

Figure 9.1 Sources and functional properties of major food hydrocolloids. Classification of major hydrocolloids based on biological origin and their primary functional roles in food systems, including thickening, gel formation, stabilization, and encapsulation.

9.10.1 Quality control, safety evaluation, and regulatory framework

9.10.1.1 Identity and purity specifications

Hydrocolloids must meet identity and purity standards to ensure predictable performance and safety. Typical specifications include viscosity, microbial limits, ash values, and heavy metals. These specifications are central to consistent industrial application.

9.10.1.2 Codex and JECFA safety framework

Internationally, the use of hydrocolloids as food additives is guided by Codex Alimentarius provisions, including the General Standard for Food Additives (GSFA), which outlines permitted use by food category and associated conditions (Codex STAN 192–1995) (Codex Alimentarius Commission, 1995). In parallel, Joint FAO/WHO Expert Committee on Food Additives (JECFA) provides detailed monographs specifying identity and purity criteria (e.g., assay requirements, limits for contaminants, and compositional parameters) that support consistent quality and safety. Collectively, these frameworks facilitate regulatory alignment, support international trade, and strengthen consumer protection through standardized specifications.

9.11 Practical selection guide for formulation

Hydrocolloid selection should be based on the intended product structure and processing conditions.

9.11.1 Selection by functional objective

- Thickening/stabilization required: guar gum, xanthan gum, gum arabic
- Gel structure required: pectin, alginate, carrageenan, gelatin

9.11.2 Selection by processing and pH

- Acidic systems → gum arabic, xanthan; pectin for fruit gels
- Cold-set gels → alginate + calcium
- Dairy stabilization → carrageenan (often with LBG)

9.11.3 Selection by consumer and label requirements

- Vegan/plant-based → pectin, alginate, carrageenan
- Premium melt-in-mouth confectionery → gelatin

9.12 Future perspectives

Emerging directions include development of optimized hydrocolloid blends for sugar reduction, texture engineering in plant-based dairy analogs, and polymer-based encapsulation platforms for probiotics and micronutrients. Clean-label demand is expected to further stimulate innovation in natural gum processing and functionality enhancement, particularly in nutraceutical foods (Phillips and Williams, 2021).

9.13 Conclusion

Natural gums and gelling agents are indispensable hydrocolloids in food processing due to their capacity to control viscosity, stabilize multiphase systems, and create structured gels. Their functional behavior is governed by polymer chemistry, ionic characteristics, and interactions with food constituents, making matrix-specific selection essential. Synergistic blends provide superior texture and stability compared with single hydrocolloids and have become standard in industrial applications. Codex and JECFA frameworks provide globally recognized standards that support safe and standardized use of these additives in modern foods.

AI disclosure

During the preparation of this work, the authors used Open AI for Grammar correction. After using this tool/service, the authors reviewed and edited the content as needed and take full responsibility for the content of the publication.

References

Codex Alimentarius Commission. General Standard for Food Additives (GSFA). Codex STAN 192-1995 (CXS 192-1995). Rome: FAO/WHO; 1995 (updated).

De Vries, J., Rombouts, F.M., Voragen, A.G.J., Pilnik, W., 2002. The effect of carrageenan on the stability of dairy systems: Interactions with casein micelles. Food Hydrocolloids 16 (2), 131–141. Available at https://doi.org/10.1016/S0268-005X(01)00058-3.

Draget, K.I., Smidsrød, O., Skjåk-Bræk, G., 2005. Alginates from algae. In: Steinbüchel, A., Rhee, S.K. (Eds.), Polysaccharides and Polyamides in the Food Industry. Wiley-VCH, Weinheim, pp. 1–30.

Ellis, P.R., Rayment, P., Wang, Q., 1996. A physico-chemical perspective of plant polysaccharides in relation to glucose absorption, insulin secretion and the entero-insular axis. The Proceedings of the Nutrition Society 55, 881–898.

Garcia-Ochoa, F., Santos, V.E., Casas, J.A., Gomez, E., 2000. Xanthan gum: Production, recovery, and properties. Biotechnology Advances 18 (7), 549–579. Available at https://doi.org/10.1016/S0734-9750(00)00050-1.

Grant, G.T., Morris, E.R., Rees, D.A., Smith, P.J.C., Thom, D., 1973. Biological interactions between polysaccharides and divalent cations: The egg-box model. FEBS Letters 32 (1), 195–198. Available at https://doi.org/10.1016/0014-5793(73)80770-7.

Imeson, A. (Ed.), 2009. Food stabilisers, thickeners and gelling agents. Blackwell Publishing Ltd, Oxford ISBN: 978-1-405-13267-1.

Nishinari, K., 2009. Some thoughts on the definition of a gel. Progress in Colloid & Polymer Science 136, 87–94. Available at https://doi.org/10.1007/978-3-540-92803-4_15.

Mortensen, A., Aguilar, F., Crebelli, R., Di Domenico, A., Frutos, M. J., Galtier, P., Gott, D., Gundert-Remy, U., Lambré, C., Leblanc, J.-C., Lindtner, O., Moldeus, P., Mosesso, P., Oskarsson, A., Parent-Massin, D., Stankovic, I., Waalkens-Berendsen, I., Woutersen, R. A., Wright, M., Younes, M., Brimer, L., Peters, P., Wiesner, J., Christodoulidou, A., Lodi, F., Tard, A., & Dusemund, B. (2017). Scientific opinion on the re-evaluation of guar gum (E 412) as a food additive. EFSA Journal, 15(2), Article 4669. https://doi.org/10.2903/j.efsa.2017.4669

Phillips, G.O., Williams, P.A. (Eds.), 2021. Handbook of hydrocolloids, 3rd ed. Woodhead Publishing/Elsevier, Cambridge ISBN: 978-0-12-820104-6.

Rinaudo, M., 2008. Main properties and current applications of some polysaccharides as biomaterials. Polymer International 57 (3), 397–430. Available at https://doi.org/10.1002/pi.2378.

Saha, D., Bhattacharya, S., 2010. Hydrocolloids as thickening and gelling agents in food: A critical review. Journal of Food Science and Technology 7 (6), 587–597. Available at https://doi.org/10.1007/s13197-010-0162-6.

Sriamornsak, P. (2007). Chemistry of Pectin and Its Pharmaceutical Uses : A Review.

Thakur, B.R., Singh, R.K., Handa, A.K., 1997. Chemistry and uses of pectin—A review. Critical Reviews in Food Science and Nutrition 37 (1), 47–73. Available at https://doi.org/10.1080/10408399709527767.

Voragen, A.G.J., Coenen, G.J., Verhoef, R.P., Schols, H.A., 2009. Pectin, a versatile polysaccharide present in plant cell walls. Structural Chemistry 20 (2), 263–275. Available at https://doi.org/10.1007/s11224-009-9442-z.

Acidity regulators: sources and benefits

10

Umalatha[1], *Shilpa P.*[2], *and M.R. Raghunath*[1]
[1]Department of Food Science and Technology, Padmashree Institute of Management and Sciences, Bengaluru, Karnataka, India, [2]Department of Nutrition and Dietetics, Padmashree Institute of Management and Sciences, Bengaluru, Karnataka, India

10.1 Introduction

Acidity regulators constitute an important class of food additives that are used to control, maintain, or modify the acidity or alkalinity of foods. Acidity is a fundamental chemical property of foods that influences not only taste but also microbial stability, enzyme activity, texture, and overall shelf life. In food processing, even minor changes in pH can significantly alter product quality and safety.

Acidity regulators are therefore added deliberately during processing to ensure that foods remain within an optimal pH range required for stability, palatability, and regulatory compliance. Unlike preservatives that primarily exert direct antimicrobial effects, acidity regulators modify the food environment itself, thereby indirectly influencing microbial growth and chemical reactions (Damodaran et al., 2007).

10.1.1 Definition of acidity regulators

Acidity regulators are substances that are added to foods for the purpose of controlling or adjusting pH. They may increase acidity, reduce acidity, or stabilize pH at a desired level throughout processing and storage.

According to the Codex Alimentarius Commission, acidity regulators include acids, bases, and buffering agents that modify or control the acidity or alkalinity of foods and beverages (Codex Alimentarius Commission. General Standard for Food Additives GSFA, 2023).

Similarly, the Food Safety and Standards Authority of India recognize acidity regulators as approved food additives used to achieve technological functions related to pH control under specified limits and food categories (Food Safety and Standards Authority of India FSSAI, 2011).

Thus, acidity regulators may function as:

- Acidulants, which lower pH
- Alkalizers, which raise pH
- Buffers, which resist changes in pH

Nutraceutical Applications in Food Processing and Preservation. DOI: https://doi.org/10.1016/B978-0-443-45784-5.00005-8

10.1.2 Purpose of using acidity regulators

The primary purpose of acidity regulators is to maintain an appropriate pH that supports product stability, safety, and sensory quality. pH affects numerous food system parameters, including microbial growth, enzyme activity, protein functionality, and flavor perception.

In beverages, acidity regulators help achieve refreshing sourness and microbial stability. In bakery and dairy products, they influence dough behavior, protein coagulation, and texture. In processed foods, pH control is essential for consistency across batches and compliance with food safety standards (Branen et al., 2002).

From a technological perspective, acidity regulators act as chemical control points, ensuring that foods remain within a predictable and functional pH window throughout their shelf life.

10.2 Functional role of acidity regulators

The functional significance of acidity regulators in food systems extends far beyond simple sourness adjustment. By controlling the hydrogen ion concentration of foods, these additives influence a wide range of chemical, biochemical, microbiological, and sensory processes (Belitz et al., 2009). As pH is a master variable in food chemistry, acidity regulators act as key technological tools for ensuring product stability, safety, and uniform quality during processing and storage (Damodaran et al., 2007).

The major functional roles of acidity regulators can be broadly categorized into pH control, flavor balance, and shelf-life extension.

10.2.1 pH control

pH control is the primary and most critical function of acidity regulators. Many food components—such as proteins, enzymes, pigments, and polysaccharides—are highly sensitive to changes in pH. Even small deviations can lead to undesirable effects such as protein denaturation, color changes, textural defects, or loss of microbial stability.

Acidity regulators are therefore used to:

- Adjust pH to a desired level during formulation
- Maintain pH consistency across batches
- Stabilize pH during processing, storage, and distribution

For example, in dairy products, controlled acidification governs milk protein coagulation and texture development, while in beverages, pH adjustment ensures clarity, stability, and microbial safety. In processed foods, maintaining a consistent pH is essential for reproducible quality and regulatory compliance (Branen et al., 2002).

Thus, acidity regulators function as chemical stabilizers, anchoring foods within an optimal pH range required for functionality.

10.2.2 Flavor balance

Acidity is one of the five basic taste sensations and plays a crucial role in shaping flavor perception. Acidity regulators influence not only sourness but also the overall balance of sweetness, saltiness, and bitterness in foods.

Organic acids such as citric, lactic, and malic acids contribute characteristic sour notes that enhance freshness and palatability, particularly in beverages, fruit products, and confectionery. At controlled levels, acidity can suppress excessive sweetness, enhance fruity flavors, and mask undesirable aftertastes (Damodaran et al., 2007).

In complex food systems, acidity regulators are used to fine-tune flavor profiles rather than simply increase acidity. For instance, buffered acid systems allow gradual acid release, providing a smoother taste perception and avoiding harsh sourness.

From a sensory standpoint, acidity regulators act as flavor modulators, ensuring harmony among different taste components.

10.2.3 Shelf-life extension

Acidity regulators indirectly contribute to shelf-life extension by creating conditions unfavorable for microbial growth and chemical spoilage. Many spoilage and pathogenic microorganisms exhibit limited growth at low pH values (Davidson et al., 2005). By lowering or stabilizing pH, acidity regulators enhance the effectiveness of preservation strategies without acting as direct antimicrobial agents.

In acidified foods such as fruit juices, sauces, pickles, and ready-to-eat products, pH control is a critical safety parameter. Proper acidification reduces reliance on high levels of chemical preservatives and supports product safety during extended storage (Branen et al., 2002).

Additionally, acidity regulators slow down enzymatic browning, oxidative reactions, and pigment degradation, thereby preserving color, texture, and nutritional quality. In this way, they function as indirect shelf-life enhancers, complementing preservatives and packaging systems.

10.3 Mode of action

The technological effectiveness of acidity regulators is rooted in their ability to control the hydrogen ion concentration of foods. By modifying pH, these additives influence chemical equilibria, enzyme activity, microbial growth, and the physical behavior of food components. The mode of action of acidity regulators can be broadly explained through three fundamental mechanisms: acidification, neutralization, and buffering (Damodaran et al., 2007). Understanding these mechanisms is essential for selecting appropriate acidity regulators and applying them effectively in different food systems (Fig. 10.1).

10.3.1 Acidification

Acidification refers to the intentional lowering of pH by adding acidulants. Acidulants dissociate in aqueous systems to release hydrogen ions (H^+), thereby

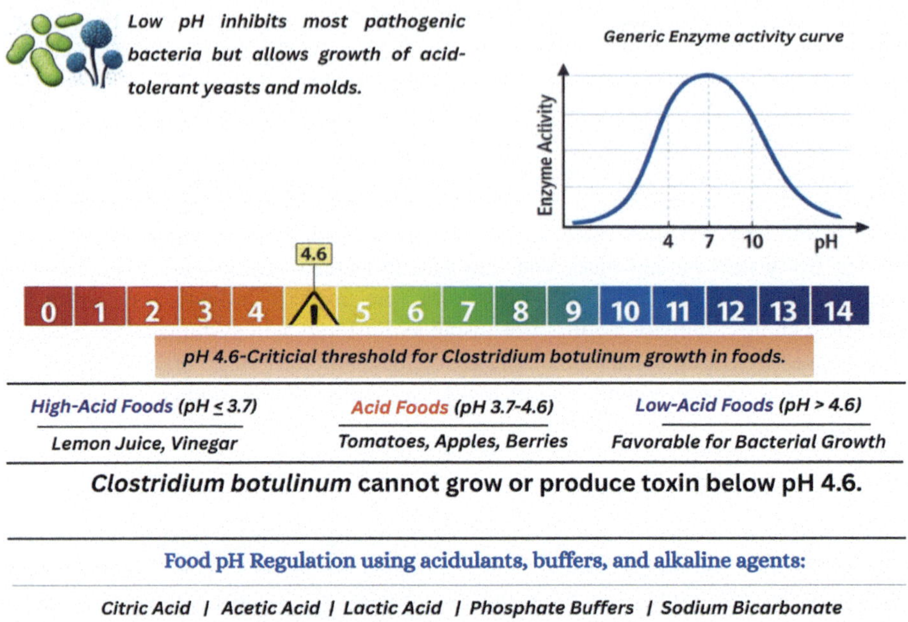

Figure 10.1 Mode of action of acidity regulators in food systems. Illustration of the three primary mechanisms by which acidity regulators function in foods: acidification through hydrogen ion release, neutralization via acid–base reactions, and buffering systems that stabilize pH.

increasing acidity. The effectiveness of acidulants is influenced by their pKa, which determines the proportion of undissociated acid that contributes to antimicrobial activity.

Organic acids such as citric acid, lactic acid, acetic acid, and malic acid are commonly used as acidulants in foods. Their effectiveness depends on factors such as dissociation constant (pKa), solubility, and interaction with other food components (Belitz et al., 2009). Weak organic acids are particularly effective because their undissociated form can penetrate microbial cell membranes, contributing to microbial inhibition in acidic foods (Branen et al., 2002).

In food processing, acidification is used to achieve microbial safety in acidified foods, control enzymatic activity, influence protein coagulation and gel formation, and enhance flavor and freshness perception.

For example, acidification plays a critical role in fermented dairy products, fruit beverages, sauces, and pickled foods, where pH reduction ensures safety and characteristic sensory properties (Damodaran et al., 2007).

10.3.2 Neutralization

Neutralization involves the adjustment of excessive acidity by adding alkaline substances that react with acids to form salts and water. This process raises pH and reduces sharp or undesirable sourness.

Alkalizing agents such as sodium bicarbonate, potassium bicarbonate, and calcium carbonate are commonly used in bakery and confectionery applications. In bakery products, neutralization is particularly important for balancing the acidity generated by acidulants or fermentation, thereby influencing dough behavior, gas release, and final product texture (Branen et al., 2002).

Neutralization also helps to prevent excessive protein denaturation, by reducing the corrosive effects of high acidity and improving palatability in acidic formulations. Thus, neutralization serves as a corrective mechanism, ensuring that acidity remains within a functional and sensory acceptable range.

10.3.3 Buffering

Buffering refers to the ability of a system to resist changes in pH when acids or bases are added. Buffering agents typically consist of weak acids and their corresponding salts or weak bases and their salts.

Common buffering agents in foods include sodium citrate, potassium citrate, sodium phosphate, and lactate systems. These compounds stabilize pH through equilibrium between weak acids and their conjugate bases, allowing controlled release or uptake of hydrogen ions (Damodaran et al., 2007).

Buffer systems are particularly important in foods where a stable pH is required for enzyme activity, maintaining flavor consistency and stability during processing, which involves heat or mechanical stress. For instance, in processed cheese and beverages, buffering systems prevent sudden pH fluctuations that could otherwise lead to texture defects, protein precipitation, or flavor instability. From a functional perspective, buffering acts as a pH shock absorber, protecting the food matrix from abrupt chemical changes.

10.3.4 Integrated mode of action

In practical food formulations, acidification, neutralization, and buffering often operate simultaneously rather than independently. A beverage may be acidified for flavor and safety, buffered for pH stability, and partially neutralized to avoid harsh sourness. This integrated approach allows precise pH control while optimizing sensory and functional properties.

Thus, the mode of action of acidity regulators reflects a balance between chemical control and sensory design, making them indispensable tools in food processing and formulation.

10.4 Types and classification of acidity regulators

Acidity regulators encompass a diverse group of substances that differ in their chemical nature, functional behavior, and application in food systems. For clarity and practical relevance, these additives are commonly classified based on their functional role in modifying pH, rather than solely on chemical composition. This classification reflects how acidity regulators are selected and applied during food formulation and processing. Based on their primary function, acidity regulators are broadly classified into acidulants, alkalizers, and buffering agents (Branen et al., 2002; Damodaran et al., 2007) (Fig. 10.2).

10.4.1 Acidulants

Acidulants are substances added to foods to increase acidity by lowering pH. They are among the most widely used acidity regulators due to their multiple functional roles, including flavor enhancement, microbial control, and processing stability.

Common food acidulants include citric acid, lactic acid, acetic acid, malic acid, tartaric acid, and phosphoric acid. These acids differ in strength, dissociation behavior, and sensory characteristics, allowing formulators to tailor acidity profiles according to product requirements. Acidulants are used to enhance sour taste and freshness, improve microbial stability in acidified foods, control enzymatic activity and facilitate protein coagulation and gel formation.

For example, citric acid is widely used in beverages and fruit products for its clean, refreshing sourness, while lactic acid contributes a milder acidity typical of fermented foods. Thus, acidulants serve both technological and sensory functions in food systems (Damodaran et al., 2007).

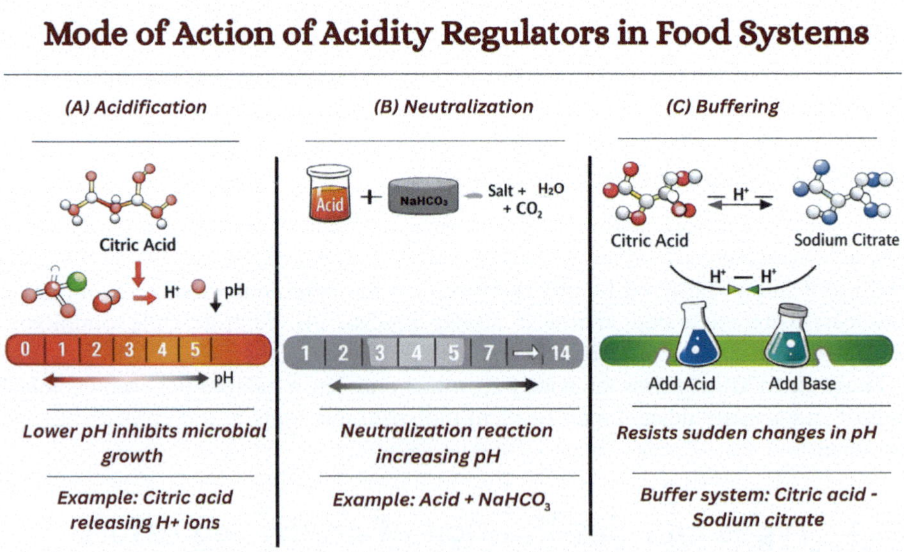

Figure 10.2 Food pH scale and microbial safety threshold in foods. Food pH scale showing microbial growth limitations, enzyme activity dependence on pH, and the critical safety threshold of pH 4.6 below which *Clostridium botulinum* cannot grow. The figure also illustrates classification of foods based on acidity

10.4.2 Alkalizers

Alkalizers are substances used to reduce acidity or increase pH in food products. They counteract excessive acidification and help achieve a balanced pH that supports optimal processing and sensory quality. Common alkalizing agents include sodium bicarbonate, potassium bicarbonate, calcium carbonate, and ammonium bicarbonate. These compounds react with acids to form salts, water, and, in some cases, carbon dioxide.

In bakery products, alkalizers play a critical role by neutralizing acids produced during fermentation, facilitating leavening reactions and influencing crumb structure and texture. Excessive acidity in foods can lead to harsh taste, protein destablization, and processing challenges. Alkalizers, therefore, act as corrective agents, ensuring that pH remains within a functional range suitable for both processing and consumption (Branen et al., 2002).

10.4.3 Buffering agents

Buffering agents are compounds that resist changes in pH, even when acids or bases are added. They typically consist of weak acids and their salts or weak bases and their salts, forming buffer systems that stabilize pH. Common food buffers include sodium citrate, potassium citrate, sodium phosphate, potassium phosphate, and lactate systems. These compounds maintain pH by reversible binding or release of hydrogen ions, thereby preventing sudden pH fluctuations during processing and storage (Damodaran et al., 2007).

Buffering agents are particularly important in foods where pH stability is essential for texture and flavor, heat treatment or storage could alter acidity, and enzyme activity must be carefully controlled. For instance, in processed cheese and beverage formulations, buffering systems ensure consistent texture, prevent protein precipitation, and maintain flavor balance throughout shelf life.

Although acidulants, alkalizers, and buffering agents are classified separately, they are often used in combination within a single formulation. A food product may be acidified to ensure safety, buffered to maintain pH stability, and partially neutralized to achieve acceptable sensory quality. This integrated approach highlights that the classification of acidity regulators is functional rather than rigid, reflecting the complex chemical and sensory requirements of modern food systems.

10.5 Natural acidity regulators

Natural acidity regulators are substances derived from biological sources that are used to adjust or stabilize the pH of foods. With increasing consumer preference for naturally sourced ingredients, these regulators have gained prominence in modern food formulations. Beyond consumer acceptance, natural acidity regulators are valued for their compatibility with food matrices, mild sensory profiles, and multifunctional roles in flavor development and preservation. Natural acidity regulators are commonly obtained from organic acids, fermentation processes, and plant-based sources, each contributing distinct functional and sensory characteristics (Damodaran et al., 2007).

10.5.1 Organic acids

Organic acids are the most widely used natural acidity regulators in food systems. They are weak acids that partially dissociate in aqueous environments, allowing controlled acidification and buffering action.

Common organic acids used as acidity regulators include citric acid, lactic acid, malic acid, tartaric acid, and acetic acid. These acids occur naturally in fruits, fermented foods, and biological tissues (Belitz et al., 2009). Citric acid, for example, is abundant in citrus fruits and is extensively used in beverages and fruit-based products due to its clean, refreshing sourness. Malic acid, naturally present in apples, provides a smoother and more persistent acidity, while tartaric acid is characteristic of grapes and wine products (Damodaran et al., 2007).

From a functional standpoint, organic acids lower pH effectively, enhance microbial stability, improve flavor brightness, and support enzymatic and protein-based reactions. Their weak acid nature allows better control over acidity compared to strong mineral acids, making them particularly suitable for food applications.

10.5.2 Fermentation-derived acids

Fermentation-derived acids are produced through microbial metabolism of carbohydrates and are integral to many traditional and modern food products. These acids not only regulate acidity but also contribute to characteristic flavors and textures.

Key fermentation acids include lactic acid (produced by lactic acid bacteria), acetic acid (produced by acetic acid bacteria), and gluconic acid (produced during oxidative fermentation).

In fermented dairy products, lactic acid formation lowers pH gradually, facilitating protein coagulation and enhancing microbial safety. In vinegar and fermented condiments, acetic acid provides both acidity and preservative action. Fermentation-derived acids offer the advantage of gradual acidification, which results in smoother sensory perception and improved textural development compared to direct acid addition (Branen et al., 2002).

10.5.3 Plant-based sources of acidity regulators

Several plant-derived ingredients function as natural acidity regulators due to their inherent organic acid content. Fruit juices, concentrations, and plant extracts are commonly used to adjust pH while contributing to natural flavor and color.

Examples include, lemon and lime juice concentrates, tamarind pulp, amla (Indian gooseberry) extract, grape, and apple concentrates. These sources are particularly relevant in traditional foods and clean-label formulations, where ingredient familiarity is important. In addition to acidity regulation, plant-based sources may provide bioactive compounds such as polyphenols and vitamins, enhancing the nutritional profile of foods. However, variability in acid composition and concentration requires careful standardization to ensure consistent pH control during processing and storage (Damodaran et al., 2007).

Natural acidity regulators offer a combination of technological functionality, sensory compatibility, and consumer acceptance. Their ability to regulate pH while contributing desirable flavor attributes makes them especially valuable in beverages, dairy products, fermented foods, and fruit-based formulations.

Nevertheless, effective use of natural acidity regulators requires precise formulation and quality control to account for natural variability and ensure product consistency.

10.6 Synthetic acidity regulators

Synthetic acidity regulators are chemically manufactured substances used to control or modify the acidity of foods with high precision and consistency. Unlike naturally derived regulators, synthetic acidity regulators are valued primarily for their predictable chemical behavior, high purity, and cost efficiency, which make them particularly suitable for large-scale food processing.

These regulators are widely used in industrial food systems where strict pH control, uniformity across batches, and extended shelf life are essential. Synthetic acidity regulators can be broadly categorized into mineral acids and synthetic salts based on their chemical nature and functional role (Branen et al., 2002).

10.6.1 Mineral acids

Mineral acids are inorganic acids that dissociate readily in aqueous systems, releasing hydrogen ions and thereby lowering pH rapidly. Common mineral acids used as acidity regulators in food processing include phosphoric acid, hydrochloric acid and sulfuric acid, which are used mainly as processing aids under controlled conditions rather than as direct food additives.

Among these, phosphoric acid is the most widely used mineral acid in foods, particularly in carbonated beverages, processed cheeses, and certain bakery applications. It provides strong and consistent acidification and contributes a sharp acidic taste profile (Damodaran et al., 2007).

Mineral acids are typically used when rapid and effective pH reduction is required, strong buffering systems are present in the food matrix and flavor contribution from organic acids is not desired. Due to their high dissociation and strength, mineral acids must be carefully controlled to avoid excessive acidity, corrosion of processing equipment, or undesirable sensory effects.

10.6.2 Synthetic salts

Synthetic salts are produced by neutralizing acids with bases and are commonly used as buffering agents or acidity stabilizers in food systems. These salts do not impart strong acidity but help maintain pH within a narrow and desirable range.

Common synthetic salts used as acidity regulators include sodium citrate, potassium citrate, sodium phosphate, calcium phosphate and sodium lactate. These compounds are particularly important in foods where pH stability is critical for texture, flavor, and processing behavior. For example, sodium citrate and phosphate salts are widely used in processed cheese to stabilize protein dispersion and prevent phase separation during heating (Damodaran et al., 2007).

Synthetic salts function by providing buffering capacity, preventing abrupt pH fluctuations and enhancing functional properties of proteins and emulsions. Their controlled composition and reproducibility make them indispensable in standardized food formulations.

10.6.3 Functional advantages of synthetic acidity regulators

Synthetic acidity regulators offer several technological advantages, including high purity and consistent performance, greater stability under heat and storage conditions, precise pH control at low usage levels and cost-effectiveness in large-scale production. These attributes make them particularly suitable for beverages, processed foods, dairy analogs, and convenience foods where consistency and shelf stability are paramount (Branen et al., 2002). However, their synthetic origin may influence consumer perception, especially in clean-label products, necessitating careful communication and regulatory compliance.

From a food technology standpoint, synthetic acidity regulators serve as precision tools that enable tight control over pH-dependent processes. While they may lack the additional sensory or bioactive benefits of natural regulators, their reliability and efficiency ensure consistent product quality and safety across industrial food systems.

10.7 Natural versus synthetic acidity regulators

The choice between natural and synthetic acidity regulators is an important formulation decision in modern food systems. While both categories serve the fundamental purpose of pH control, they differ in terms of source, safety perception, cost, performance, and consumer acceptance. Understanding these differences allows food technologists to select appropriate regulators based on product type, processing conditions, regulatory requirements, and market expectations.

10.7.1 Source origin

Natural acidity regulators are derived from biological sources such as fruits, fermented substrates, and plant materials. Examples include citric acid from citrus fruits, lactic acid from fermentation, and organic acids naturally present in fruit concentrates. Their origin is often emphasized in product labeling and marketing, particularly in clean-label foods.

In contrast, synthetic acidity regulators are produced through controlled chemical processes. These include mineral acids such as phosphoric acid and chemically synthesized salts such as sodium citrate and phosphate salts. Although synthetically produced, many of these compounds are chemically identical to their naturally occurring counterparts (Damodaran et al., 2007).

Thus, the distinction between natural and synthetic regulators is based on production method rather than chemical functionality.

10.7.2 Safety profile

From a scientific and regulatory perspective, both natural and synthetic acidity regulators are considered safe when used within prescribed limits. Safety evaluation is based on toxicological data, metabolic fate, and long-term exposure assessment, irrespective of whether the additive is natural or synthetic.

International regulatory bodies such as the Codex Alimentarius Commission and national authorities like the Food Safety and Standards Authority of India approve acidity regulators based on scientific risk assessment rather than source origin (Codex Alimentarius Commission. General Standard for Food Additives GSFA, 2023; Food Safety and Standards Authority of India FSSAI, 2011).

Therefore, the perception that natural regulators are inherently safer than synthetic ones is not always scientifically justified; safety depends on dose, exposure, and regulatory compliance, not origin alone (Branen et al., 2002).

10.7.3 Cost and functional performance

Synthetic acidity regulators generally offer advantages in terms of cost efficiency, purity, and batch-to-batch consistency. Their predictable chemical behavior allows precise pH control, making them particularly suitable for large-scale industrial processing.

Natural acidity regulators, while often preferred for label appeal, may exhibit variability in composition due to differences in raw material source, season, and processing conditions. This variability can affect acid strength and buffering capacity, requiring additional standardization steps (Damodaran et al., 2007).

As a result, synthetic regulators are often favored in products where tight pH control and cost optimization are critical, whereas natural regulators are preferred when sensory contribution and label positioning are prioritized.

10.7.4 Consumer acceptance

Consumer perception plays a significant role in the selection of acidity regulators. Natural acidity regulators are generally viewed more favorably due to their association with traditional foods and minimal processing. This preference has intensified with the rise of clean-label and "natural" food trends.

Synthetic acidity regulators, despite their safety and functionality, may face resistance due to unfamiliar chemical names or misconceptions regarding synthetic additives. Consequently, manufacturers often balance functional requirements with consumer expectations by combining natural acids with buffering salts or by using nature-identical compounds declared appropriately on labels (Branen et al., 2002).

- Natural acidity regulators offer better consumer acceptance and sensory compatibility but may present challenges in consistency and cost.
- Synthetic acidity regulators provide precision, stability, and economic advantages but may require careful communication to address consumer perception.

An informed selection of acidity regulators, therefore, requires integration of scientific evidence, technological needs, regulatory compliance, and consumer expectations, rather than reliance on source origin alone.

10.8 Nutraceutical benefits

In addition to their primary technological functions, certain acidity regulators contribute beneficial physiological effects when consumed as part of a balanced diet. These effects are generally supportive rather than therapeutic and arise from the interaction of organic acids with digestive processes, mineral bioavailability, microbial ecology, and oxidative stability of foods.

While acidity regulators are not nutraceuticals in a strict sense, several naturally occurring acids exhibit nutraceutical relevance due to their role in promoting digestive efficiency, nutrient absorption, and food safety (Damodaran et al., 2007).

10.8.1 Digestive health

Organic acids may support digestive processes by influencing gastric conditions and enhancing enzyme activity, although primary control of gastric acidity is physiological. Adequate acidity supports the activity of digestive enzymes, particularly pepsin, and facilitates protein digestion.

Fermentation-derived acids, especially lactic acid, are associated with fermented foods that promote gut health by supporting beneficial microbial populations. Lower gastric pH also serves as a natural barrier against ingested pathogens, thereby indirectly supporting gastrointestinal health (Branen et al., 2002).

Thus, acidity regulators present in fermented and acidified foods can aid digestive processes by enhancing enzyme efficiency and microbial balance.

10.8.2 Mineral absorption

Certain organic acids enhance the bioavailability of dietary minerals by forming soluble complexes that improve intestinal absorption. Citric acid is particularly effective in chelating minerals such as calcium, iron, and magnesium, keeping them in a soluble form suitable for absorption in the intestine.

For example, citrate complexes improve calcium solubility in dairy and fortified foods, supporting mineral uptake without altering sensory quality. This property is especially relevant in populations with marginal mineral intake or increased nutritional requirements (Damodaran et al., 2007). Thus, acidity regulators can play an indirect but important role in improving nutritional quality through enhanced mineral bioavailability.

10.8.3 Antimicrobial action

Although many acidity regulators are not primarily classified as preservatives, they contribute significantly to microbial inhibition through pH reduction. Many spoilage and pathogenic microorganisms exhibit reduced growth or survival at low pH values.

Organic acids exert antimicrobial effects through their undissociated form, which penetrates microbial cell membranes and disrupts intracellular pH homeostasis (Davidson et al., 2005). This mechanism enhances food safety and complements other preservation strategies (Branen et al., 2002). In this context, acidity regulators support food safety while reducing the need for higher levels of chemical preservatives.

10.8.4 Antioxidant support

Certain acidity regulators contribute indirectly to antioxidant protection by stabilizing pH conditions that minimize oxidative reactions. Acidic environments slow down enzymatic browning and lipid oxidation, thereby preserving color, flavor, and nutrient integrity.

Additionally, organic acids such as citric acid act as metal chelators, binding pro-oxidant metal ions like iron and copper that catalyze oxidative reactions. This chelating action enhances the effectiveness of antioxidants and helps maintain food quality during storage (Damodaran et al., 2007). Thus, acidity regulators support antioxidant systems by creating favorable chemical conditions rather than acting as antioxidants themselves.

From a nutraceutical viewpoint, acidity regulators support digestive enzyme activity, improve mineral solubility and absorption, enhance microbial safety, and stabilize antioxidant systems in foods.

These benefits emphasize their multifunctional relevance in food systems, extending beyond technological roles to supportive nutritional functions.

10.9 Food applications of acidity regulators

Acidity regulators are widely used across diverse food categories due to their ability to control pH, enhance sensory quality, improve processing stability, and support food safety. Their application is guided by the specific functional requirements of each food system, as pH influences texture, flavor, microbial stability, and shelf life in different ways.

The major food applications of acidity regulators include beverages, bakery foods, dairy products, and processed foods.

10.9.1 Beverages

Beverages represent one of the largest application areas for acidity regulators. pH control in beverages is essential for flavor balance, microbial stability, and clarity. Organic acids such as citric, malic, and phosphoric acids are commonly used to provide refreshing sourness, enhance fruit flavor perception, stabilize color and aroma compounds, and inhibit microbial growth.

In carbonated soft drinks, phosphoric or citric acid contributes to characteristic taste profiles while maintaining product stability. In fruit juices and ready-to-drink beverages, acidity regulators help standardize pH despite natural variation in raw materials, ensuring consistent quality across batches (Damodaran et al., 2007).

10.9.2 Bakery foods

In bakery products, acidity regulators play a crucial role in dough development, leavening reactions, and final product texture. Alkalizers such as sodium bicarbonate react with acidulants to produce carbon dioxide, which contributes to dough expansion and crumb structure.

Acidity regulators in bakery applications are used to control fermentation rate, optimize gas release during baking, improve crumb texture and volume, and prevent excessive browning or off flavors. Buffer systems are also employed to maintain optimal pH for yeast activity and enzyme function, ensuring consistent product quality (Branen et al., 2002).

10.9.3 Dairy products

Dairy foods are highly sensitive to pH changes due to the behavior of milk proteins and minerals. Acidity regulators are therefore essential in controlling coagulation, texture, and stability in dairy products.

In fermented dairy products such as yoghurt and cultured milk, gradual acidification through lactic acid formation governs protein gelation and flavor development. In processed cheese, buffering salts such as sodium citrate and phosphate salts stabilize emulsions and prevent phase separation during heating (Damodaran et al., 2007). Thus, acidity regulators help achieve desired texture, meltability, and shelf stability in dairy systems.

10.9.4 Processed and convenience foods

In processed and ready-to-eat foods, acidity regulators support both safety and sensory quality. These products often undergo thermal processing and extended storage, making pH control essential.

Acidity regulators are used in sauces, ketchup, dressings, pickles, chutneys, canned foods, retort-processed foods, and ready-to-eat products. In such foods, acidity regulators help maintain microbial safety, stabilize color and flavor, and ensure regulatory compliance for acidified food products. By controlling pH, they also enhance the effectiveness of preservatives and packaging systems (Branen et al., 2002).

Across food categories, acidity regulators function as versatile formulation tools that adapt to the specific chemical and sensory requirements of each product. Their appropriate selection and controlled use enable manufacturers to deliver safe, stable, and palatable foods while meeting regulatory and consumer expectations.

10.10 Safety regulations

The use of acidity regulators in foods is governed by stringent safety regulations to ensure that these additives do not pose health risks when consumed over long periods. Because acidity regulators directly influence a critical determinant of food safety, microbial stability, and chemical reactivity, their use is subject to rigorous scientific evaluation and regulatory oversight.

Regulatory control ensures that acidity regulators are technologically justified, toxicologically safe, and used within clearly defined limits appropriate for specific food categories.

10.10.1 Permissible limits and acceptable daily intake

Safety evaluation of acidity regulators is based on toxicological studies that assess their metabolic fate, potential toxicity, and long-term exposure effects. For many organic acids and buffering salts, safety assessments have established that they are metabolized as part of normal physiological pathways.

International expert committees determine:

- NOAEL (No Observed Adverse Effect Level) from animal and human studies
- ADI (Acceptable Daily Intake), defined as the amount of an additive that can be consumed daily over a lifetime without appreciable health risk

For several commonly used acidity regulators, such as citric acid, lactic acid, and sodium citrate, regulatory authorities have concluded that intake at levels used in foods does not present a safety concern when used according to good manufacturing practices (GMP) (WHO & FAO, 1987).

Maximum permissible levels for acidity regulators are specified based on the type of food product, technological need and expected dietary exposure. These limits ensure consumer safety while allowing effective pH control.

10.10.2 International regulatory bodies

At the international level, the Codex Alimentarius Commission plays a central role in harmonizing food additive regulations. The Codex General Standard for Food Additives (GSFA) lists approved acidity regulators along with their INS numbers, permitted food categories, and maximum usage levels (Codex Alimentarius Commission. General Standard for Food Additives GSFA, 2023).

Scientific risk assessment supporting Codex decisions is conducted by the Joint FAO/WHO Expert Committee on Food Additives (JECFA). JECFA evaluates toxicological data and establishes ADI values that form the basis of national regulations.

10.10.3 National regulatory frameworks

In India, acidity regulators are regulated by the Food Safety and Standards Authority of India under the Food Safety and Standards Act, 2006. The Food Safety and

Standards (Food Products Standards and Food Additives) Regulations, 2011, specify approved acidity regulators, food categories in which they may be used, maximum permissible levels and labeling requirements. These regulations are periodically updated to reflect advances in scientific knowledge and international harmonization (Food Safety and Standards Authority of India FSSAI, 2011).

Similarly, other national authorities such as the European Food Safety Authority (EFSA) and the US Food and Drug Administration (FDA) regulate acidity regulators based on scientific risk assessment and continuous re-evaluation.

10.10.4 Labeling and compliance

Regulatory frameworks mandate transparent labeling of acidity regulators to inform consumers and facilitate compliance. Acidity regulators must be declared either by their specific name or by their functional class, followed by the INS number, for example, acidity regulator (citric acid) and acidity regulator (INS 330).

Such labeling ensures transparency, supports informed consumer choice, and aids regulatory monitoring. Overall, safety regulations governing acidity regulators are designed to ensure that only scientifically evaluated substances are permitted, additives are used at technologically justified levels, and consumer exposure remains within safe limits. Through coordinated international and national oversight, acidity regulators continue to play a safe and essential role in modern food systems.

10.11 Clean label trends in the use of acidity regulators

In recent years, the food industry has witnessed a pronounced shift toward clean-label formulations, driven by increasing consumer demand for transparency, simplicity, and perceived naturalness in food ingredients. Clean-label foods are generally characterized by shorter ingredient lists, recognizable components, and minimal use of additives perceived as synthetic or artificial.

This trend has significantly influenced the selection and declaration of acidity regulators in food products, encouraging greater reliance on naturally derived acids and fermentation-based ingredients.

10.11.1 Preference for natural acidity regulators

Consumers increasingly associate natural acidity regulators—such as citric acid from citrus fruits, lactic acid from fermentation, and plant-based acid sources—with traditional food processing and health. As a result, food manufacturers are reformulating products to replace synthetic-sounding additives with naturally sourced or nature-identical alternatives.

In beverages, dairy products, and fruit-based foods, organic acids and fermentation-derived acids are often preferred because they provide pH control while contributing to familiar sensory profiles. Such ingredients align well with clean-label expectations and support positive consumer perception (Baines and Seal, 2012).

However, it is important to note that the chemical function of an acidity regulator remains unchanged regardless of source. Thus, the preference for natural regulators is largely driven by consumer perception rather than functional superiority.

10.11.2 Industry shift and formulation strategies

The clean-label movement has prompted the food industry to adopt several strategic approaches, including the replacement of synthetic acids with organic or fermentation-derived acids, the use of ingredient blends that provide acidity and buffering effects naturally, and reformulation to reduce additive dependence through process optimization

At the same time, manufacturers must ensure that reformulation does not compromise product safety, stability, or sensory quality. Natural acidity regulators may exhibit variability in strength and composition, necessitating additional quality control measures to maintain consistent pH and shelf life (Damodaran et al., 2007).

10.11.3 Balancing clean label and food safety

While clean-label initiatives reflect evolving consumer values, it is essential to emphasize that food safety remains nonnegotiable. Acidity regulators play a critical role in microbial control, particularly in acidified and ready-to-eat foods. Eliminating or inadequately replacing effective acidity regulators in the pursuit of clean labels may increase the risk of spoilage or foodborne illness.

Regulatory authorities and food scientists consistently emphasize that additive selection must be guided by scientific risk assessment and technological need, rather than perception alone (Branen et al., 2002). Therefore, clean-label reformulation should be supported by robust validation studies to ensure that safety and quality standards are maintained.

10.11.4 Future outlook

The future of acidity regulators lies in achieving a balance between clean-label expectations and technological functionality. Continued research into fermentation-derived acids, plant-based extracts, and innovative processing technologies may help reduce reliance on conventional additives while preserving food safety and quality.

Transparent communication, accurate labeling, and consumer education will also play a key role in aligning public perception with scientific evidence. Ultimately, clean-label trends should complement—not replace—science-based food formulation practices.

Acidity regulators remain indispensable components of modern food systems, functioning as critical tools for pH control, sensory optimization, safety assurance, and shelf-life extension. As consumer expectations evolve, the challenge for food technologists lies in integrating clean-label principles with regulatory compliance and scientific rigor.

A balanced, evidence-based approach ensures that acidity regulators continue to support both technological innovation and public health protection (Table 10.1).

Table 10.1 Classification of acidity regulators as per FSSAI/Codex terminology, their functions, and limitations.

Regulatory category	**Name (Codex/FSSAI-recognized examples)**	**Primary technological function**	**Limitations/ regulatory considerations**
Acidulants (organic acids)	Citric acid (INS 330), lactic acid (INS 270), acetic acid (INS 260), malic acid (INS 296), tartaric acid (INS 334)	Reduction and control of pH, flavor enhancement, inhibition of microbial growth	Excessive acidity may affect sensory acceptability; usage restricted to GMP or specified limits
Acidulants (inorganic acids)	Phosphoric acid (INS 338)	Strong acidification, buffering support particularly in beverages	Subject to maximum permitted levels; concerns regarding mineral balance at high intake
Buffering agents / acidity regulators (salts of acids)	Sodium citrate (INS 331), potassium citrate (INS 332), calcium citrate (INS 333), sodium lactate (INS 325)	pH stabilization, buffering action, mineral supplementation	Sodium- or potassium-related dietary concerns; restricted use in specific food categories
Alkalizing agents	Sodium bicarbonate (INS 500), potassium bicarbonate (INS 501)	Neutralization of acidity, leavening action in bakery products	Overuse may result in off-flavors, textural defects, and elevated sodium levels
Phosphates (acidity regulators / emulsifying salts)	Sodium phosphate (INS 339), potassium phosphate (INS 340), calcium phosphate (INS 341)	Buffering, emulsification, moisture retention, mineral balance	Increasing regulatory scrutiny; limits imposed due to phosphorus intake concerns
Acidity regulators with delayed action	Glucono-δ-lactone (INS 575)	Gradual acidification, texture development in dairy and tofu products	Limited consumer familiarity; classified as synthetic by clean-label standards
Naturally derived acidulants	Vinegar (INS 260), citrus-derived citric acid, fermentation-derived lactic acid	Mild acidification, clean-label positioning, flavor modulation	Lower acid strength; variability in composition; limited buffering capacity

Acidity regulators permitted under FSSAI and Codex Alimentarius are broadly classified as acidulants, buffering agents, alkalizing agents, and phosphates based on their technological function in food systems. Organic acidulants such as citric, lactic, and acetic acids remain the most widely used due to their multifunctional roles in pH control, flavor enhancement, and microbial stability. Buffering salts and phosphates provide pH stability and functional advantages but are subject to regulatory limits owing to mineral intake considerations. Recent regulatory and consumer-driven trends increasingly favor naturally derived acidity regulators, particularly fermentation-based acids, to meet clean-label expectations while maintaining product safety and quality.

AI disclosure

During the preparation of this work, the authors used OpenAI and Copilot for the preparation of this manuscript. ChatGPT (OpenAI) was utilized for assistance in refining and organizing the text. Microsoft Copilot was used to generate illustrative images for conceptual visualization. After using this tool/service, the authors reviewed and edited the content as needed and take full responsibility for the content of the publication.

References

Baines, D., Seal, R., 2012. Natural food additives, ingredients and flavourings natural food additives, ingredients and flavourings. Elsevier Ltd, United Kingdom, pp. 1–460. Available at http://www.sciencedirect.com/science/book/9781845698119.

Belitz, H.D., Grosch, W., Schieberle, P., 2009. Food chemistry. Springer Available at https://doi.org/10.1007/978-3-540-69934-7.

Branen, A.L., Davidson, P.M., Salminen, S., Thorngate, J.H. 2002. Food Additives. 2nd ed. CRC Press. Available at https://doi.org/10.1201/9780203904030.

Codex Alimentarius Commission. General Standard for Food Additives (GSFA), 2023. FAO/WHO. Available at https://www.fao.org/fao-who-codexalimentarius/codex-texts/list-standards/en/.

Davidson, P.M., Sofos, J.N., Branen, A.L., 2005. Antimicrobials in food. CRC Press Available at https://doi.org/10.1201/9781420028737.

Damodaran, S., Parkin, K.L., Fennema, O.R. 2007. Fennema's Food Chemistry. 4th ed. CRC Press. https://doi.org/10.1201/9781420027976.

Food Safety and Standards Authority of India (FSSAI), 2011. Food products standards and food additives regulations. (as amended). Available at https://www.fssai.gov.in/cms/food-safety-and-standards-regulations.php.

WHO & FAO, 1987. Principles for the safety assessment of food additives and contaminants in food. Environmental Health Criteria 70. Available at https://apps.who.int/iris/handle/10665/37287.

Natural emulsifiers

11

Nagamani J.E.[1] *and Sudipta Kumar Mohanty*[2]
[1]Department of Food Science and Nutrition, Padmashree Institute of Management and Sciences, Bengaluru, Karnataka, India, [2]Department of Biotechnology, Padmashree Institute of Management and Sciences, Bengaluru, Karnataka, India

11.1 Introduction

Emulsifiers are a class of amphiphilic molecules possessing both hydrophilic and lipophilic domains allowing them to localize at the interface between immiscible phases such as oil and water. This unique structure reduces the interfacial tension facilitating the formation and stabilization of emulsions systems in which one liquid is dispersed in another (McClements and Gumus, 2016).

In food systems, emulsifiers play a vital role in products where the coexistence of oil and water phases is essential for the structure and quality of mayonnaise, salad dressings, margarine, ice cream, and baked goods. Emulsifiers contribute to improved stability against phase separation, droplet reduction and enhanced dispersion. They also modify texture, support aeration in foams, and improve the overall sensory perception of food products (Dickinson, 2011).

11.1.1 Difference between emulsifiers, stabilizers, and surfactants

Although emulsifiers, stabilizers, and surfactants are sometimes used interchangeably, they differ in scope and function:

11.1.1.1 Emulsifiers

- Emulsifiers promote the formation of oil-in-water (O/W) or water-in-oil (W/O) emulsions.
- They lower the interfacial tension and help maintain a uniform dispersion of droplets.
- For example, lecithin in chocolate prevents fat bloom and improves the smoothness.

11.1.1.2 Stabilizers

- Stabilizers increase the viscosity of the continuous phase and physically slow down the droplet movement and coalescence.
- They prevent destabilization by forming a protective network around droplets.
- Common examples include Xanthan gum, guar gum, carrageenan, etc.

11.1.1.3 Surfactants

- Surfactants are surface active agents that reduce interfacial tension, modify foaming, improve spreadability, or enhance fat dispersion.
- Includes detergents, wetting agents, foaming agents, and emulsifiers.

Nutraceutical Applications in Food Processing and Preservation. DOI: https://doi.org/10.1016/B978-0-443-45784-5.00010-1

Some of the commonly used surfactants are monoglycerides and diglycerides, polysorbates, sucrose esters of fatty acids, etc.

11.1.2 Importance in modern food formulations

Emulsifiers enhance rheology influencing the texture, mouthfeel, shelf life, and consumer acceptance. In addition, they improve aeration in batters, control fat crystallization, and support creamy or spreadable textures. Modern food manufacturing industries use emulsifiers for technological benefits like product reformulation and innovation including the use of reduced fat, dispersion and stability of bioactive compounds like fat-soluble vitamins, antioxidants, and flavors, enhancing their delivery in complex food systems (Garti and Leser, 2001).

11.1.2.1 Stabilization of emulsions

Emulsifiers form a protective layer around dispersed droplets preventing their aggregation, thereby maintaining the stability, uniform texture and better appearance. Without emulsifiers, emulsions are prone to creaming, coalescence, or flocculation, which negatively affect product quality and consumer acceptance.

11.1.2.2 Influence on texture and mouthfeel

Emulsifiers significantly influence the sensory properties of food by modulating the size and distribution of fat droplets impacting the creaminess, smoothness, and perceived richness. In bakery products, emulsifiers such as lecithin or mono- and diglycerides interact with starch and protein molecules to improve dough stability and enhance crumb softness (Garti and Leser, 2001). Similarly, in dairy and nondairy beverages, emulsifiers contribute to a uniform, velvety mouthfeel, improving palatability and overall consumer satisfaction.

11.1.2.3 Enhancement of shelf-life

Emulsifiers contribute indirectly to the shelf life of food products by maintaining droplet integrity, slow down lipid oxidation and prevent the production of off flavors. Additionally, emulsifiers work synergistically with other stabilizers, antioxidants, and hydrocolloids to prolong the physical and sensory stability of processed foods (McClements and Gumus, 2016).

11.1.2.4 Optimizing consumer acceptance

Emulsifiers ensure product uniformity, stability, promote smoothness, creaminess, and consistent flavor release thereby contribute to the desirable sensory attributes (Dickinson, 2011). In baked and dairy products, they improve volume, softness, and shelf life. Furthermore their inclusion enables the reformulation of low-fat and plant-based foods without compromising sensory appeal satisfying the health conscious consumers.

11.1.3 Shift from synthetic to natural emulsifiers

The food industry is shifting from synthetic to natural emulsifiers due to consumer demand for sustainability, health benefits, and clean label transparency. Though synthetic emulsifiers like mono- and diglycerides or polysorbates are efficient and cost effective, concerns over their origin and impact on health have driven interest in natural alternatives. Natural emulsifiers lecithin, proteins, saponins, and polysaccharides offer multifunctional properties providing additional nutritional or functional benefits, such as antioxidant activity or improved digestibility. Consumer perception of natural as healthier has prompted brands to reformulate products without compromising texture, taste, or shelf life (Aspri et al., 2020). Clean label trends have also influenced regulatory and marketing practices, encouraging simpler ingredient lists and transparent labeling.

11.1.4 Scope of the chapter

This chapter explores the diverse sources of plant-, animal-, and microbial-derived emulsifying agents, their structural characteristics, and functional properties, emphasizing their mechanisms of action. Further this section extends the role of emulsifiers in improving texture, shelf life, and sensory quality of food formulations while aligning with clean-label and sustainability trends. Moreover, the chapter examines the nutraceutical potential of natural emulsifiers, focusing on their bioactive properties and contributions to health promoting foods.

11.2 Fundamentals of emulsions in food

An emulsion is a thermodynamically unstable, heterogeneous system consisting of two immiscible liquids typically oil and water in which one liquid is dispersed as droplets within the other, stabilized by surface-active agents or emulsifiers (McClements and Gumus, 2016). The dispersed phase forms droplets that are prevented from coalescing by emulsifiers, which reduce interfacial tension and form protective interfacial films (Tadros, 2013).

11.2.1 Types of emulsions

Depending on the composition and the nature of the continuous and dispersed phases, emulsions are classified into three major types: oil-in-water (O/W),water-in-oil (W/O), and multiple emulsions such as W/O/W and O/W/O systems.

11.2.1.1 Oil-in-water) emulsions

Oil-in-water (O/W) emulsions consist of oil droplets dispersed in a continuous aqueous phase. Hydrophilic emulsifiers such as lecithin, casein, whey proteins, Tween 20, and Tween 80 stabilize O/W emulsions by reducing interfacial tension and preventing droplet coalescence through electrostatic or steric mechanisms. These emulsions enhance flavor encapsulation, texture, and uniform fat dispersion while improving sensory appeal and shelf life stability (McClements and Gumus, 2016).

11.2.1.2 Water-in-oil emulsions

Water-in-oil (W/O) emulsions, or reverse emulsions, consist of water droplets dispersed in a continuous oil phase. Sorbitan esters, glycerol monostearate, and fatty acid derivatives emulsions offer low conductivity, greasy, hydrophobic texture that helps to retain moisture and creamy texture, stabilize flavors, and encapsulate water to enhance spreadability (Dickinson, 2011).

11.2.1.3 Multiple emulsions (W/O/W and O/W/O)

Multiple emulsions are more complex systems in which droplets of one liquid phase contain smaller droplets of another immiscible phase. These systems are classified mainly into water-in-oil-in-water (W/O/W) and oil-in-water-in-oil (O/W/O) emulsions.

11.2.1.3.1 Water-in-oil-in-water (W/O/W) emulsions

Water-in-oil-in-water (W/O/W) multiple emulsions consist of water droplets encapsulated within oil globules, which are further dispersed in an external aqueous phase. W/O/W emulsions are used for controlled release of bioactives like vitamins, minerals, and drugs, improving bioavailability, also enabling low-fat foods to mimic the creamy texture of high-fat products while enhancing nutritional profiles (Rayner et al., 2014).

11.2.1.3.2 Oil-in-water-in-oil (O/W/O) emulsions

Oil-in-water-in-oil (O/W/O) emulsions contain water droplets with internal oil droplets, all dispersed in an external oil phase, providing both hydrophilic and lipophilic environments for ingredient delivery. They are used in confectionery, coatings, and fat-rich foods to enhance texture and sensory appeal (McClements and Gumus, 2016).

11.2.2 Factors determining emulsion types

Emulsion type oil-in-water (O/W) or water-in-oil (W/O) depends on the balance of physicochemical and mechanical factors affecting droplet size, interfacial behavior, and phase distribution. Its formation is governed by thermodynamic and kinetic interactions among the dispersed and continuous phases, emulsifiers, and processing conditions (Tadros, 2013; McClements and Gumus, 2016). Understanding these factors is essential for designing stable emulsions in food, pharmaceutical, and cosmetic applications.

11.2.2.1 Nature and ratio of phases

The relative volume fraction of oil and water phases is a critical determinant of emulsion formation. Typically, the phase present in higher concentration tends to become the continuous phase, while the phase present in smaller proportion forms dispersed droplets. A higher oil-to-water ratio favors the formation of water-in-oil (W/O) emulsions, whereas an excess aqueous phase promotes oil-in-water (O/W) emulsions (Dickinson, 2011). Phase inversion may occur when the ratio exceeds a critical limit or when temperature and emulsifier composition shift during processing.

11.2.2.2 Type and concentration of emulsifier

The hydrophilic lipophilic balance (HLB) of the surfactant determines its preference for either oil or water. Surfactants with high HLB values (8–18) promote O/W emulsions while those with low HLB values (3–6) favor W/O emulsions (Tadros, 2013). The emulsifier concentration must be optimized; insufficient amounts lead to droplet aggregation, while excessive surfactant may induce phase inversion or foaming (McClements and Gumus, 2016).

11.2.2.3 Viscosity of the phases

High viscosity continuous phases hinder droplet movement thereby enhancing stability. However excessively viscous systems can also limit droplet disruption during homogenization, leading to larger droplet sizes and reduced emulsion uniformity. Thus, balancing viscosity is essential for maintaining desirable rheological properties and stability.

11.2.2.4 Interfacial properties and droplet size

Lower interfacial tension facilitates smaller droplet formation, improving kinetic stability by reducing gravitational separation and coalescence (Dickinson, 2011). Smaller droplets also possess greater surface area, allowing more efficient surfactant adsorption. The droplet size distribution determines the optical appearance, mouthfeel, and physical stability of emulsions (McClements and Gumus, 2016).

11.2.2.5 Temperature and phase inversion

Temperature changes viscosity, surface tension, and how well surfactants dissolve. For nonionic surfactants, the phase inversion temperature (PIT) is the point where the surfactant shifts from hydrophilic to lipophilic behavior. Emulsions made close to this temperature usually have small droplets and good stability, but sudden temperature changes can cause the emulsion to break or invert (Tadros, 2013).

11.2.2.6 Method and intensity of agitation

High shear homogenization produces finer droplets and stable O/W emulsions, while insufficient agitation yields coarse, unstable systems (McClements and Gumus, 2016). The energy density and duration of emulsification influence droplet coalescence, surfactant adsorption rate, and the final microstructure of the emulsion.

11.2.2.7 Additives and stabilizing agents

Biopolymers like pectin, gum arabic, and xanthan gum improve physical stability by reducing droplet mobility and retarding creaming (Rayner et al., 2014). In addition, pickering particles composed of natural biopolymers or food-grade minerals offer enhanced stability through irreversible adsorption at the oil–water interface (Dickinson, 2011).

11.2.2.8 Environmental factors

External conditions such as pH, ionic strength, and storage temperature influence emulsion stability. Changes in pH can modify the surface charge and zeta potential, affecting electrostatic repulsion between droplets. Similarly, high salt concentrations can cause charge screening, leading to aggregation or flocculation. Effective regulation of environmental factors is therefore critical to maintain emulsion integrity.

11.3 Classification of natural emulsifiers

In recent years, the interest in natural emulsifiers has increased significantly due to consumer demand for clean-label products, biodegradability, and environmental sustainability. Natural emulsifiers are widely distributed in nature and can be broadly classified into plant-based, animal-based, microbial-derived, and marine-derived emulsifiers. Each source group possess unique chemical structures that govern their emulsifying performance and stability characteristics relevant to the industrial applications (Dickinson, 2011) (Fig. 11.1).

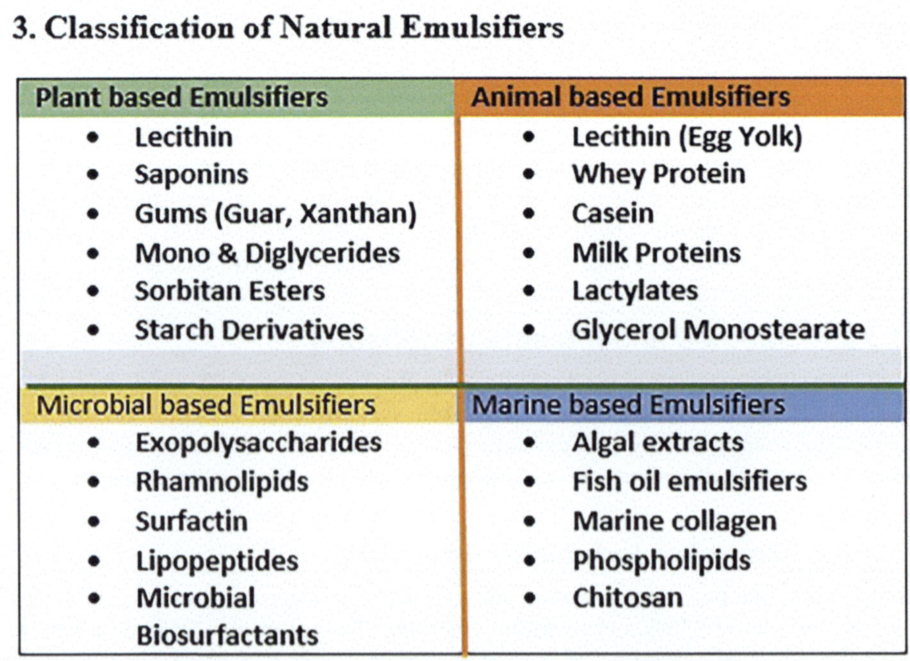

Figure 11.1 Classification of emulsifiers based on natural sources.

11.3.1 Plant-based emulsifiers

Plants serve as sustainable sources of natural emulsifiers namely phospholipids, hydrocolloids, saponins, and proteins often preferred as Generally Recognized as Safe (GRAS) substances.

- **Phospholipids (lecithin):** Derived primarily from soybean, sunflower, or rapeseed. Lecithins are complex mixtures of phosphatidylcholine, phosphatidylethanolamine, and phosphatidylinositol. They reduce interfacial tension and form stable oil-in-water emulsions.
- **Hydrocolloids:** Natural gums such as gum arabic, pectin, guar gum, and xanthan gum are plant polysaccharides that stabilize emulsions by increasing viscosity and providing steric stabilization.
- **Saponins:** These are naturally occurring amphiphilic glycosides composed of hydrophobic triterpenoid backbone and hydrophilic sugar moieties found in plants like *Quillaja saponaria* and *Yucca schidigera*).
- **Plant proteins:** Proteins derived from soybean, pea, and lentil exhibit emulsifying capabilities due to their ability to unfold and form a viscoelastic film that prevents droplet coalescence to adsorb at the oil–water interface.

11.3.2 Animal-based emulsifiers

Animal-derived emulsifiers are primarily composed of proteins, phospholipids, and waxes.

- **Egg yolk lecithin:** Rich in phospholipids (especially phosphatidylcholine) and lipoproteins, commonly used in the preparation of mayonnaise, creams, and sauces.
- **Milk proteins:** Caseins and whey proteins act as efficient natural emulsifiers due to their amphiphilic amino acid composition. They form flexible interfacial films that resist coalescence and form heat stable emulsions.
- **Beeswax and lanolin:** Beeswax and lanolin are hydrophobic components used in cosmetic and pharmaceutical formulations as stabilizers and structuring agents.

11.3.3 Microbial-derived emulsifiers

Microorganisms-derived emulsifiers include glycolipids, lipopeptides, phospholipids, and polymeric compounds.

- **Rhamnolipids,** produced through microbial fermentation by *Pseudomonas aeruginosa*, consist of rhamnose sugar units linked to hydroxy fatty acids. Rhamnolipids improve the dough stability and batter texture and decrease hardness and chewiness of bakery products.
- **Sophorolipids**, synthesized by nonpathogenic yeasts, mainly Candida bombicola, consist of a sophorose disaccharide linked to a fatty acid chain. As emulsifiers they improve the texture and stability of the products.
- **Mannosylerythritol lipids (MELs)** are glycolipid surfactants mainly produced by yeasts belonging to the genus *Pseudozyma* and *Ustilago* species. MELs consist of a hydrophilic sugar alcohol head and hydrophobic fatty acid tails contributing to the mouthfeel and creaminess in low fat or fat reduced formulations.
- **Trehalolipids** are other notable biosurfactants produced by actinobacteria such as *Rhodococcus*, *Nocardia*, and *Mycobacterium* species. These compounds are composed of a trehalose disaccharide linked to one or two long-chain fatty acids. They are used to stabilize food emulsions such as salad dressings, sauces, and dairy-based beverages.

11.3.4 Marine-derived emulsifiers

Emulsifiers obtained from marine organisms such as algae, fish, and marine microorganisms include algal polysaccharides and phospholipids.

- **Algal polysaccharides** such as alginate, extracted from brown seaweeds (*Laminaria*, *Ascophyllum*, *Macrocystis*), consist of β-D mannuronic and α-L guluronic acid residues. They form gel like networks enhancing the viscosity and emulsion stability. Fucoidan and laminarin, sulfated and nonsulfated polysaccharides, respectively, contribute to emulsion stability also offering antioxidant and antimicrobial benefits .
- **Marine phospholipids**, derived from krill oil and fish roe, contain high proportions of omega-3 fatty acids such as eicosapentaenoic acid (EPA) and docosahexaenoic acid (DHA) esterified to phosphatidylcholine and phosphatidylethanolamine. Their amphiphilic structure facilitates the formation of stable nanoemulsions, improves lipid bioavailability, and contributes to the sensory quality of emulsified food products.

11.3.5 Functional category of natural emulsifiers

Natural emulsifiers stabilize the emulsions through interfacial adsorption, electrostatic repulsion, steric hindrance, and viscosity enhancement, which influence stability, texture, and mouthfeel. Based on their molecular properties and behavior at the oil–water interface, they can be classified as listed in Table 11.1.

11.4 Extraction of natural emulsifiers

The extraction process is largely governed by the nature of the source material and the specific class of emulsifiers involved such as phospholipids, proteins, polysaccharides, and glycolipids each requiring tailored conditions to preserve their structural integrity and functional activity (McClements and Gumus, 2016).

11.4.1 Phospholipid emulsifiers

Phospholipids from oilseeds such as soybean, sunflower, rapeseed, and corn are mainly recovered during degumming. Water or dilute acids are used to hydrate phospholipids and separate as gums. Solvent extraction using ethanol or hexane–ethanol mixtures enhances the purity. Phospholipids extraction from animal sources including egg yolk and milk uses organic solvents and membrane concentration. Further purification is achieved by chromatography. Microbial and marine phospholipids require cell disruption followed by solvent extraction and chromatography. Sustainable methods such as enzyme-assisted extraction and supercritical CO_2 improve selectivity and reduce solvent residues.

11.4.2 Polysaccharide emulsifiers

Plant gums and mucilages are extracted by soaking or refluxing powdered material in water at 70°C, followed by filtration and alcohol precipitation. Pectin is obtained through acidic hot-water extraction and alcohol precipitation. Microwave, ultrasound,

Table 11.1 Functional classification of natural emulsifiers.

Functional category	Mechanism of action	Examples
Monomolecular film-formers	These emulsifiers lower interfacial tension, and form a stable monomolecular film around dispersed droplets.	Phospholipids, saponins, microbial glycolipids
Macromolecular stabilizers	These form thick, viscoelastic interfacial layers that prevent droplet coalescence and enhance kinetic stability.	Milk proteins (casein, whey), soy protein, gum arabic, pectin
Particulate or pickering stabilizers	These produce a physical steric barrier that prevents droplet coalescence without reducing interfacial tension.	Cellulose nanocrystals, starch granules, chitosan particles, protein aggregates
Thickening and structuring agents	These increase the viscosity of the continuous phase, restrict droplet movement, and stabilize emulsions through entanglement or gel formation.	Xanthan gum, guar gum, carrageenan, alginate
Surface-active and structuring agents	These emulsifiers exhibit multiple mechanisms namely reducing interfacial tension while forming interfacial networks or increasing viscosity.	Lecithin-protein blends, polysaccharide–protein complexes

and enzyme-assisted methods enhance yield and preserve structure. Algal polysaccharides are recovered using hot water or mild alkaline treatments, followed by filtration and alcohol precipitation. Further purification using ultrafiltration or ion-exchange chromatography is applied to obtain better purity.

11.4.3 Protein emulsifiers

Plant proteins from soybeans, chickpeas, and peas are extracted by alkaline solubilization at a pH of 8–10 followed by isoelectric precipitation (pH 4.5–5.0) and ultrafiltration. Further spray drying helps improve the yield. Dairy and egg proteins are recovered using pH adjustment or controlled precipitation. Marine and microalgal proteins require enzymatic hydrolysis and cell disruption further purified using membrane filtration or isoelectric precipitation. Ultrasound and enzyme-supported methods further enhance solubility and preserve emulsifying functionality.

11.4.4 Saponin emulsifiers

Saponins, the amphiphilic glycosides are extracted by refluxing with solvents, followed by liquid–liquid partitioning or column chromatography. For example, in *Asparagus,* to align with green chemistry principles, supercritical CO_2 extraction, pressurized hot-water extraction, and microwave-assisted extraction have been developed as efficient methods.

11.4.5 Extraction of animal-derived waxes

Marine sources such as fish and krill provide wax esters and lipoprotein like emulsifiers. These compounds are typically extracted using ethanol or isopropanol reflux, followed by liquid–liquid partitioning to remove triglycerides. Supercritical CO_2 extraction has emerged as a cleaner method, offering selective recovery of wax esters at controlled pressures and temperatures without leaving solvent residues.

11.4.6 Extraction of microbial-derived emulsifiers

Microorganisms such as bacteria, yeasts, and fungi produce emulsifiers like glycolipids, lipopeptides, and polymeric compounds. Glycolipids are recovered via centrifugation followed by acid precipitation and solvent extraction (Banat et al., 2010). Lipopeptides like surfactin, iturin, and fengycin are isolated by pH precipitation, followed by solvent extraction, ultrafiltration, and chromatography.

High molecular weight polymeric bioemulsifiers such as emulsan, alasan, liposan, and biodispersan are recovered by mild cell disruption followed by solvent precipitation using alcohol. Subsequent purification steps include dialysis, ultrafiltration, or gel filtration chromatography to remove low molecular weight impurities.

11.4.7 Lipid-based emulsifiers extraction from marine organisms

Marine lipid emulsifiers from microalgae, seaweeds, and microbial biomass are efficiently extracted using organic solvents like chloroform–methanol systems, further followed by liquid–liquid partitioning, solid-phase extraction, or chromatographic fractionation to separate glycolipids from neutral lipids and pigments (Harwood and Guschina, 2009).

Advanced techniques such as supercritical carbon dioxide extraction with ethanol as a cosolvent and ultrasound-assisted and pressurized liquid extraction further increase the lipid yield from the marine sources (Chemat et al., 2019) (Table 11.2).

11.5 Health benefits of bioemulsifiers

Plant-derived emulsifiers like lecithins, proteins, and polysaccharides help in lipid digestion and nutrient absorption. Phosphatidylcholine, as acetylcholine precursor, supports liver health, prevents fat accumulation, and maintains cognitive and neural functions. Gum arabic and pectin act as dietary fibers to promote gut microbiota, boost short-chain fatty acid production and supports immune health (Phillips and Phillips, 2011).

Table 11.2 Overview of natural emulsifiers in food system.

Emulsifier	Source	Chemical nature	Role in food	Nutraceutical benefits	Stability notes
Soy lecithin	Soybean	Phospholipids	O/W emulsions, antispattering	Cognitive, liver health	Sensitive to oxidation
Gum arabic	Acacia tree	Polysaccharide–protein complex	Flavor encapsulation, beverage stability	Prebiotic effect	Stable over wide pH
Casein	Milk	Amphiphilic protein	Dairy emulsions	Bioactive peptides	Heat-stable
Saponins	Quillaja bark	Glycosides	Foaming, emulsifying	Cholesterol-lowering	Bitter at high levels
Xanthan gum	Microbial	Exopolysaccharide	Viscosity, stabilization	Gut health	pH-stable

Sphingomyelin and phosphatidylserine support cognitive development and gut barrier integrity and improve immunity in infants and the elderly (Brink et al., 2020). Egg lecithin reduces cholesterol absorption, promotes HDL formation, and supports cardiovascular health .

Marine extracted phospholipids are rich in omega-3 fatty acids and provide antiinflammatory, cardioprotective, and neuroprotective effects while enhancing lipid metabolism (Burri et al., 2012).

Microbial-derived emulsifiers exhibit antimicrobial and antibiofilm activities that help prevent infections in oral, skin, and gastrointestinal systems (Banat et al., 2010). Sophorolipids and MELs display anticancer effects by inducing apoptosis and inhibiting cancer cell migration (Andrade et al., 2022). Additionally, microbial biosurfactants modulate immune responses as well as chronic inflammation .

11.6 Stability and performance factors of bioemulsifiers

The functional performance and long-term stability of natural emulsifiers in food, pharmaceutical, and cosmetic systems are governed by several physicochemical and formulation-related factors. A clear understanding of these parameters is essential for developing the stable emulsions (McClements and Gumus, 2016).

11.6.1 pH and ionic strength effects

pH and ionic strength strongly influence the interfacial behavior of proteins and charged polysaccharides. Variations in pH alter the ionization state and molecular conformation, affecting electrostatic repulsion between droplets and overall emulsion stability. Protein-based emulsifiers are especially weak near their isoelectric point where reduced surface charge can lead to aggregation and flocculation. Ionic strength further modulates stability by screening electrostatic interactions and droplet coalescence (McClements and Gumus, 2016).

11.6.2 Temperature stability and heat processing tolerance

Heat treatment can induce protein unfolding, which may enhance surface activity but also increase susceptibility to aggregation and emulsion breakdown. Phospholipid-based emulsifiers generally exhibit good thermal tolerance but prolonged heating may promote lipid oxidation. Nonetheless polysaccharide emulsifiers show high resistance to thermal degradation and are therefore widely used in pasteurized and sterilized formulations (Phillips and Phillips, 2011).

11.6.3 Compatibility with other ingredients

The performance of natural emulsifiers is closely linked to their interactions with other formulation components like electrolytes, proteins, and polyphenols. Electrolytes can modify interfacial charge density influencing droplet interactions and stability. Protein–polysaccharide interactions often result in synergistic stabilization through electrostatic complex formation or multilayer interfacial structures. However, polyphenols may bind to proteins via hydrogen bonding or hydrophobic interactions altering interfacial properties and reducing emulsifying efficiency (McClements and Gumus, 2016).

11.6.4 Role of particle size and droplet distribution

Droplet size and the distribution determine the emulsion stability, appearance, and sensory attributes. Smaller droplets increase interfacial area and reduce gravitational separation process such as creaming and sedimentation. A narrow droplet size distribution further enhances stability. Ability of natural emulsifiers to form fine droplets during homogenization depends on their adsorption kinetics and interfacial film strength.

11.6.5 Encapsulation technologies for natural emulsifiers

Techniques such as spray drying, freeze drying, liposomal encapsulation, and nanoemulsion-based delivery systems protect emulsifiers from environmental stress factors like heat, pH fluctuations, and oxidation (Augustin and Hemar, 2009). Encapsulation also enhances dispersibility, extends shelf life, and enables controlled release thereby broadening the application potential of natural emulsifiers in functional foods.

11.7 Industrial applications of natural bioemulsifiers

Natural bioemulsifiers are increasingly preferred across food, nutraceutical, and functional product industries because of their biodegradability, low toxicity, consumer acceptance, as well as multifunctional health benefits.

11.7.1 Beverages for cloud stability in juices and fortified drinks

In beverages, bioemulsifiers maintain cloud stability in fruit juices, flavored waters, and fortified drinks. Plant polysaccharides like gum arabic and modified starches prevent droplet aggregation by forming steric and electrostatic barriers, ensuring the uniform appearance and sensory quality (Phillips and Phillips, 2011). In fortified drinks, bioemulsifiers enhance dispersion and absorption of fat-soluble nutrients.

11.7.2 Bakery and confectionery for texture and shelf life enhancement

In bakery products natural emulsifiers improve dough handling, crumb softness, and volume by interacting with starch and gluten networks. Lecithin and plant proteins act as dough conditioners reducing staling and moisture loss during storage and also enhance gas retention during baking resulting in improved texture and mouthfeel.

In confectionery, they regulate fat crystallization, reduce viscosity, and prevent phase separation, thereby contributing to extended shelf life and improved product consistency without depending on synthetic additives.

11.7.3 Dairy products

In ice creams, emulsifiers promote partial coalescence of fat droplets and improve air incorporation, creaminess as well as resistance to melting apart from enhancing the sensory quality. In yogurt and fermented milk drinks, polysaccharide emulsifiers improve viscosity and enhance mouthfeel by stabilizing the protein gel matrix contributing for consumer acceptance (Phillips and Phillips, 2011).

11.7.4 Margarine, mayonnaise, and salad dressings

High-fat emulsified products such as margarine, mayonnaise, and salad dressings depend on emulsifiers to maintain stability during storage and handling. Natural emulsifiers such as lecithins, saponins, and protein–polysaccharide complexes form robust interfacial films that prevent oil droplet aggregates under mechanical and thermal stress. In reduced fat formulations, bioemulsifiers help maintain texture and mouthfeel by enhancing water binding capacity and mimicking the sensory properties of fats.

11.7.5 Nutraceutical formulations for lipid-based delivery systems

Nano- and microemulsions stabilized by phospholipids, proteins, or biosurfactants enhance solubility, stability, intestinal uptake, and bioavailability of water-insoluble compounds like curcumin, resveratrol, and omega-3s (Augustin and Hemar, 2009). Encapsulation with natural emulsifiers protects degradation and enables controlled and safer targeted release of bioactive molecules (McClements and Gumus, 2016).

11.8 Safety, regulatory, and quality aspects of natural emulsifiers

The commercial use of natural emulsifiers in food and nutraceutical formulations is governed by stringent safety, regulatory, and quality frameworks to ensure consumer protection and safety.

11.8.1 Generally recognized as safe status and European Food Safety Authority approval

US Food and Drug Administration (FDA) has classified lecithin, gum arabic, pectin, and certain microbial biosurfactants as generally recognized as safe (GRAS) based on a history of safe consumption and toxicological evidence (FDA, 2022).

European Food Safety Authority (EFSA) conducts comprehensive risk assessments addressing dietary exposure, toxicology, and technological need before authorizing emulsifiers as food additives. Approved emulsifiers are designated with E-numbers, and their use is controlled by category-specific maximum limits. Natural emulsifiers with multifunctional properties may also be regulated as novel foods when their sources or production methods are new to the EU market.

11.8.2 Allergenicity concerns

Allergenicity remains a key safety consideration for emulsifiers derived from soy and egg sources. Soy lecithin- and egg-derived phospholipids may contain trace amounts of allergenic proteins challenging potential risks to sensitive individuals. Although refining lowers allergen content, regulations require labeling when emulsifiers are derived from known allergenic sources. Plant- and microbial-derived emulsifiers usually show low allergenicity but novel microbial emulsifiers must undergo allergenicity safety evaluation.

11.8.3 Labeling requirements and clean label claims

Consumer demand for transparency has strengthened the importance of accurate labeling and clean label claims. Bioemulsifiers are often preferred because they can be labeled with familiar and consumer friendly names like lecithin, gum arabic, or pectin. Food regulations require emulsifiers to be clearly listed in the ingredient list by their functional class and specific name or E-number, based on regional rules.

11.9 Analytical and quality control methods for bioemulsifiers

Strict analytical testing methods not only ensure that products meet specification limits but also help understand their stability, oil–water interfacial behavior, and composition.

11.9.1 Emulsifying Activity Index and Emulsion Stability Index

The Emulsifying Activity Index (EAI) and Emulsion Stability Index (ESI) are key measures of emulsifier performance. EAI indicates the ability of an emulsifier to adsorb at the oil–water interface, while ESI reflects the emulsion's resistance to coalescence and phase separation over time. High EAI denotes efficient emulsification, whereas high ESI indicates good stability, and both are commonly assessed by spectrophotometric monitoring of diluted emulsions.

11.9.2 Particle size distribution

Laser diffraction is the predominant technique for measuring droplet size distributions in emulsions ranging from submicron to micrometers dimensions. A narrow and small droplet size distribution is generally associated with improved physical stability as finer droplets exhibit reduced creaming velocity and a lower tendency for coalescence.

11.9.3 Zeta potential measurement for stability prediction

Zeta potential is the electrical potential of the dispersed particles and a key indicator of colloidal stability. High absolute zeta potential values reduce droplet aggregation and phase separation. It is measured by electrophoretic light scattering and helps optimize pH, ionic strength, and formulation conditions for stable emulsions.

11.9.4 Protein and phospholipid content determination

Accurate emulsifier quantification ensures purity, labeling accuracy, and regulatory compliance. Protein is typically measured by the Kjeldahl method, while HPLC and GC–MS analyze peptides, phospholipids, and fatty acids. HPLC is commonly used to separate and quantify peptides, phospholipids, and other soluble emulsifier components using suitable detection systems. GC–MS is especially effective for fatty acid profiling after derivatization, offering sensitive lipid identification.

11.10 Natural emulsifiers—future trends and innovations

The growing demand for clean label, sustainable, and multifunctional components is encouraging for the significant innovation in the development of natural emulsifiers. Advances in biotechnology, material science and green processing techniques are restructuring the applications of bioemulsifiers in food, nutraceutical, and pharmaceutical industries.

Microbial emulsifiers like rhamnolipids, sophorolipids, and mannosylerythritol lipids are produced from renewable or agroindustrial wastes. Their biodegradability and multifunctional properties support clean label formulations, while advances in metabolic engineering improve yield and cost efficiency (Banat et al., 2010).

Marine and algal sources also represent sustainable reservoirs of functional emulsifiers with unique biological properties demanding less land and freshwater. Green extraction and biorefinery approaches further enhance their commercial and environmental viability.

AI disclosure

During the preparation of this work, the authors used Open AI for correcting grammar and spell check. After using this tool/service, the authors reviewed and edited the content as needed and take full responsibility for the content of the publication.

References

Andrade, A.P.S., Sousa, J.M.C., Dantas, L.R.F., et al., 2022. Biosurfactants as anticancer agents: Glycolipids affect tumor cell viability and migration. In: Banat, I.M., Marchant, R. (Eds.), Microbial Biosurfactants: Advances and Applications. Elsevier, Amsterdam, pp. 1–12. Available at https://doi.org/10.7324/JAPS.2022.120301.

Aspri, M., Papademas, P., Tsaltas, D., 2020. Non-dairy probiotics: Technological and functional aspects. In: Liong, M.T. (Ed.), Probiotics and Prebiotics in Food, Nutrition and Health. Elsevier, Amsterdam, pp. 433–449. Available at https://doi.org/10.1111/1471-0307.12675.

Augustin, M.A., Hemar, Y., 2009. Nano- and micro-structured assemblies for encapsulation of food ingredients. In: Sagis, L.M.C. (Ed.), Food Colloids and Interface Science. Elsevier, Amsterdam, pp. 902–912. Available at https://doi.org/10.1039/B801739P.

Banat, I.M., Makkar, R.S., Cameotra, S.S., 2010. Potential commercial applications of microbial surfactants. In: Sen, R. (Ed.), Biosurfactants. Springer/Elsevier, Amsterdam, pp. 427–444. Available at https://doi.org/10.1007/s00253-010-2589-0.

Brink, L.R., Lönnerdal, B., Hernell, O., 2020. Milk fat globule membrane: Structure, composition, and physiological effects. In: McSweeney, P.L.H., Fox, P.F. (Eds.), Advanced Dairy Chemistry: Lipids. Springer/Elsevier, Amsterdam, pp. 4929–4945. Available at https://doi.org/10.3168/jds.2019-17618.

Burri, L., Hoem, N., Banni, S., 2012. Marine omega-3 phospholipids: Metabolism and biological activities. In: Kim, S.K. (Ed.), Marine Nutraceuticals. Elsevier, Amsterdam, pp. 15401–15419. Available at https://doi.org/10.3390/ijms131115401.

Chemat, F., Rombaut, N., Meullemiestre, A., Turk, M., Perino, S., Fabiano-Tixier, A.S., et al., 2019. Green food processing techniques: Preservation, transformation, and extraction. In: Galanakis, C.M. (Ed.), Sustainable Food Processing. Elsevier, Amsterdam, pp. 44–56. Available at https://doi.org/10.1016/j.ifset.2018.12.002.

Dickinson, E., 2011. Mixed biopolymers at interfaces: Competitive adsorption and multilayer structures. In: McClements, D.J. (Ed.), Food Emulsions and Foams. Elsevier, Amsterdam, pp. 1966–1983. Available at https://doi.org/10.1016/j.foodhyd.2011.02.032.

Dickinson, E., 2011. Nanoparticles and microparticles in food emulsions. In: Norton, I.T., Spyropoulos, F. (Eds.), Modern Food Colloids. Elsevier, Amsterdam, pp. 4–12. Available at https://doi.org/10.1016/j.tifs.2011.09.006.

Garti, N., Leser, M.E., 2001. Emulsifiers and stabilizers in food systems. Food Emulsions, 3rd ed. CRC Press/Elsevier, Boca Raton, FL, pp. 1–48.

Harwood, J.L., Guschina, I.A., 2009. Algal lipid metabolism and applications. In: Cohen, Z., Ratledge, C. (Eds.), Single Cell Oils. Elsevier, Amsterdam, pp. 128–136. Available at https://doi.org/10.1016/j.plipres.2009.04.002.

McClements, D.J., Gumus, C.E., 2016. Natural emulsifiers: Biosurfactants, phospholipids, biopolymers, and colloidal particles. In: McClements, D.J. (Ed.), Food Emulsions: Current Trends and Future Perspectives. Elsevier, Amsterdam, pp. 3–26. Available at https://doi.org/10.1016/j.cis.2016.03.002.

Phillips, G.O., Phillips, D.J., 2011. Biofunctional behaviour and health benefits of emulsifiers. In: Phillips, G.O., Williams, P.A. (Eds.), Handbook of Food Hydrocolloids, 2nd ed. Elsevier, Amsterdam, pp. 194–199. Available at https://doi.org/10.1016/j.foodhyd.2010.04.011.

Rayner, M., Marku, D., Eriksson, M., Sjöö, M., Dejmek, P., Wahlgren, M., 2014. Biopolymer-based Pickering emulsions and food applications. In: Dickinson, E., Leser, M.E. (Eds.), Food Colloids: Self-Assembly and Material Science. Elsevier, Amsterdam, pp. 48–62. Available at https://doi.org/10.1016/j.colsurfa.2014.03.053.

Tadros, T.F., 2013. Emulsion formation and stability. In: Tadros, T.F. (Ed.), Emulsion formation and stability. Wiley-VCH/Elsevier, Weinheim, pp. 1–150.

Further reading

Allen, T.W., Yager, S.D., Castle, L., 2012. Characterization of emulsions by laser diffraction methods. In: McClements, D.J. (Ed.), Food emulsions: Principles, practices and techniques, 3rd ed. CRC Press, Boca Raton, FL, pp. 315–323. Available at https://doi.org/10.1016/j.jfoodeng.2011.10.021.

Asioli, D., Aschemann-Witzel, J., Caputo, V., Vecchio, R., Annunziata, A., Næs, T., et al., 2017. Making sense of the clean label trends: A review of consumer food choice behavior and discussion of industry implications. In: Galanakis, C.M. (Ed.), Trends in Clean Label Food Production. Elsevier, Amsterdam, pp. 58–71. Available at https://doi.org/10.1016/j.foodres.2017.07.022.

Chalamaiah, M., Dinesh Kumar, B., Hemalatha, R., Jyothirmayi, T., 2012. Fish protein hydrolysates: Composition, bioactivity and applications. In: Toldrá, F. (Ed.), Protein By-Products: Transformation from Waste to Value. Elsevier, Amsterdam, pp. 3020–3038. Available at https://doi.org/10.1016/j.foodchem.2012.06.100.

Francis, G., Kerem, Z., Makkar, H.P.S., Becker, K., 2002. The biological action of saponins in animal systems. In: Francis, G. (Ed.), Saponins in Food and Agriculture. Elsevier, Amsterdam, pp. 587–605. Available at https://doi.org/10.1079/BJN2002725.

Friberg, S.E., Larsson, K., Sjöblom, J., 2004. Food emulsions. In: Sjöblom, J. (Ed.), Emulsions and Emulsion Stability, 2nd ed. CRC Press/Elsevier, Boca Raton, FL, pp. 1–36.

Zhang, Y., Li, S., McClements, D.J., 2021. Protein–polysaccharide conjugates: Functional properties and food applications. In: Galanakis, C.M. (Ed.), Food Structure and Functionality. Elsevier, Amsterdam, pp. 182–194. Available at https://doi.org/10.1016/j.tifs.2021.01.020.

Industry-specific applications of food additives and nutraceutical alternatives

12

M.R. Raghunath
Department of Food Science and Technology, Padmashree Institute of Management and Sciences, Bengaluru, Karnataka, India

12.1 Introduction

Food additives, by definition, are not a normal part of the food and would seldom diverse; preservation, preparation, and processing help in manufacturing, packing, packaging, transporting, or storing foods. Food additives can be derived from plants, animals, minerals, or synthesized. A great number of additives have been developed to meet the needs of industrial food processing (WHO, 2023). Additives may help in making the sensory experience of a food better; it is also quite possible that they do not affect the taste of a food in any way. In any case, food additives end up being part of most processed foods in greater or smaller quantities. Many additives find applications across a wide variety of food sectors; emulsifiers for example are used in dairy, meat, beverage, confectionery, and other products. But propellants like nitrogen have a narrow range of applications used only for dispensing of oils or syrups without foaming.

"Nutraceuticals," a word coined by Dr Stephen DeFelice in 1989 by combining "Nutrient" and "Pharmaceutical," denotes a product derived from a traditional food source that, apart from conferring basic nutritional value, also provides additional health benefits. FSSAI (2025) considers food or food components that provide physiological benefits and help in maintaining good health beyond basic nutrition as *Nutraceuticals* (Putta, 2020).

12.2 Drivers for industry-specific additive use

While food additives are not naturally present in foods where they are added, nutraceuticals are often considered as a part of foods. This makes the job of defining "nutraceuticals" rather difficult and appropriately enough, the Codex Alimentarius Commission does not have a specific, internationally accepted definition for it (CAC, 2024). The provision of specific health benefits extending beyond basic nutrition by nutraceuticals has also led to regulatory ambiguity and a rather diverse group of substances such as nutrients, dietary supplements, herbal products, and even genetically engineered foods can be considered as nutraceuticals. The explosive growth of nutraceuticals in recent years can be attributed both to the relatively relaxed regulatory regime governing their use and the increasing public awareness about health and

Nutraceutical Applications in Food Processing and Preservation. DOI: https://doi.org/10.1016/B978-0-443-45784-5.00006-X

wellness. Nutraceuticals do not generally need pre-marketing approval from regulatory authorities, being often considered as food supplements and encounter varying degrees of regulation in different countries and may come under the purview of various regulatory bodies. They are sometimes further distinguished as nutraceuticals and functional foods. Food additives, in contrast, are not inherently present in foods, but are added externally to serve specific ends. Stringent regulations govern the use of additives and manufacturers often need to prove their safety for use in foods and obtain prior approval from the respective regulatory authority in the country of use.

Viewed in this light, the possibility of using nutraceuticals as food additives emerges as an attractive proposition for the food industry. It offers twin advantages: one of deriving additional nutrition and second using a natural ingredient to serve as a "food additive." By using ingredients that are natural, recognizable and easily understood by consumers, transparency is assured achieving the "clean label" objective in a food product. In addition, using nutraceutical as additives a manufacturer avoids the cumbersome regulatory regime associated with food additives.

The possibility of using nutraceuticals as a food additives requires that we examine the various purposes for which food additives are used and explore if nutraceuticals are already in such use or if the specific purpose of adding one can be served by substituting with a nutraceutical. This could be done in two ways; by looking at the broad purposes for which food additives are used and evaluating nutraceutical alternatives for the same. Alternatively, examining if nutraceuticals available can be repurposed as food additives. *In this review, we shall use the former alternative, while approaching it from the specific sector of food industry where they are used.*

12.2.1 Beverage industry

The beverage industry uses several food additives. These would include sugar substitutes, (both nutritive and nonnutritive), preservatives, monosodium glutamate, anticaking agent, colours, emulsifiers, fat, acidulants, sodium benzoate, stabilizers, thickeners, and gelling agents.

Sugar substitutes are perhaps the most often used additives used in beverage industry, specifically in soft drinks. Acesulfame potassium (acesulfame K), high-fructose corn syrup (HFCS), aspartame, saccharin, sucralose, stevia, and cyclamate are some of the common additives in use. But HFCS, in its commercial formulations like HFCS-42 and HFCS-55, is dominant due to its lower cost and sweetness equivalent to sucrose. Sugar substitutes in beverages need stability under acidic conditions, high temperatures, and high pressures of carbonation that are usually encountered in processing of soft drinks. HFCS meets most of these criteria as do acesulfame K and sucralose but HFCS is calorie laden and linked to type II diabetes and metabolic syndrome (Bantle, 2009). Sweetener substitution by nutraceuticals in soft drinks can be most suitably addressed by steviol glycosides. It is stable up to 80°C in the pH 2–10 range and in presence of coffee and tea and also reportedly protects ascorbic acid but decomposes with organic acids at lower pH and light (Kroyer, 1999). It also does have a bitter aftertaste, variable among its component glycosides (Tian et al., 2022). Monellin, a plant protein, though much more sweeter

(3000x than sugar) and pH stable (pH 2–9), decomposes at 50°C and has not been evaluated for safety (Zeece, 2020). Thaumatin, another natural protein obtained from the Katemfe fruit (*Thaumatococcus danielli*) and reportedly safe for human consumption (EFSA, 2021), is stable from pH 2–10 and even thermoresistant (up to 70°C) in this range. Intensely sweet (3000x), it is already approved by EU, the United States, Japan, Mexico, Switzerland, and a number of other countries as GRAS with no ADI specified (Bassoli and Merlini, 2003). It is used in Japan in soft drinks, fruit juices, chewing gum, and a range of other products. While both the protein sweeteners show much promise as nutraceutical replacements in beverages, they do have a liquorice like after taste. Sweetness of thaumatin is not perceived immediately, but delayed and has a lingering aftertaste that can last up to 2 hours. This may be a hindrance to its use in beverages (Dalavi et al., 2024).

Preservation in beverage industry is usually accomplished by thermal processing in combination with cold chains and more recently by high-pressure processing. But, synthetic preservatives like benzoates, sorbates and sulfites are also used in sodas, fruit juices, carbonated drinks, wines, and fruit drinks. Although these enjoy the affordability, ease of use, and effectiveness, consumer demand for natural and clean label products has forced the manufacturers to look at natural alternatives. Natural alternatives for synthetic preservatives are citric acid, buffered vinegar, fermentation, lemon juice, and other citrus extracts. As can be inferred, citric acid, vinegar, and probiotic fermentation affect preservation by lowering the pH to inhibit spoilage bacteria and the latter also add unique favors. Essential oils and their constituents, bacteriocins, enzymes, and chitosan have been explored as nonchemical potential antimicrobials for beverage preservation (de Souza et al., 2019). These include extracts of rosemary, grapeseed, green tea, and cranberry, which have antioxidant as well as antimicrobial properties. Applications for such extracts include sodas, juices, sauces, dairy drinks, smoothies, kombucha, kefir, and such drinks. Lemon and citrus juice are traditionally used in juices and teas to create acid conditions unfavorable for microbes (Procool, 2024). Orange peel extract at 15% was used as a natural preservative in a functional beverage made from melon seed and tea stalk caffeine (Selahvarzi et al., 2021). In addition to citric acid, ascorbic acid also finds use in fruit juice beverages, but for flavor and color retention and antioxidant stability.

Acidulants are commonplace in food industry where they not only impart sour taste but also stabilize pH and carbonation characteristics, enhance and modify flavors, and act as preservatives too. Synthetic acidulants whether used singly or in combinations still lack the flavor profile of natural acidulants, which is much more complex. Consumer preferences for healthier, low-calorie, and organic drinks are propelling the growth of beverage acidulants, especially in fruit juices, soft drinks, and smoothies. With the rising focus on product innovation, manufacturers are incorporating advanced acidulants to achieve distinct taste profiles and improve drink texture (FMI, 2025). Citric acid is the major natural acidulant in beverage industry, but malic, lactic, acetic, and tartaric acids are also used. The consistency, and performance of synthetic acidulants is sometimes desired over natural ones, which also tend to be costlier.

Alcoholic fermentation, especially by yeasts, has a long and extensive history of use for preservation of beverages, as well as ensuring safety of drinks like beer in medieval times. These purposes were also served by lactic or probiotic milk fermentations. As they are well known, they will not be discussed further and will not be dealt here. Nondairy probiotic fermentation of soy milk, whey, cereal, fruit, and vegetable juices has also been developed recently (Kaur et al., 2019).

Colors play a very important role in the beverage industry. Many fruit juices are typically associated with the color of the fruits from which they are obtained. Beverages that appear to be fruit juice but contain no actual fruits such as "fruit flavored drinks" or "juice drinks" depend on colors to provide a notional satisfaction and are usually combined with synthetic, natural, or nature identical flavors. The latter too are many times synthetic or by-products such as citrus peel oils. Presently, nutraceuticals are increasingly used as natural colorants in beverages, offering both visual appeal and potential health benefits. . Both carotenoids and curcumin have found a lot of applications in food packaging applications as distinct from direct incorporation into foods (Roy et al., 2023, Silva et al., 2022). It is heartening to note that a wide range of natural food colorants have been found and are being used for food applications (Vega et al., 2023). While natural colors offer advantages of safer, healthier perception, sustainability and functional benefits due to their antioxidant and antiinflammatory effects, they are less stable than synthetic dyes, vary in hue and saturation due to their biological origin, and can be affected by pH changes and costlier. The demand and challenges of natural food colorants have been recently reviewed by Novais et al. (2022).

The antimicrobial activity of plants, specifically spices, are well known both against spoilage and pathogenic microflora. Due to their extensive diversity in terms of flavors and nutraceutical activities, many secondary metabolites of plants such as polyphenols, poly saccharides, flavonoids, terpenes, etc. have been used in plant-based drinks (Wu et al., 2023). Coffee and tea are too well known to be discussed further, but other plant extracts like turmeric, vanilla, mint, and lavender are equally well known in the beverage industry and used both for flavor enhancement, nutritional benefits, and functional properties (Arjuna Natural, 2025). Almond, soy, oat, coconut, and rice milks have become popular recently as an alternative to dairy milk. These are mostly preserved by refrigeration, freezing, heat treatment (pasteurization or sterilization), and addition of preservatives, but newer techniques such as pascalization, pulsed electric field, and ohmic heating are also employed (Bocker and Silva, 2022). Herbal infusions (tisanes) from tulsi, chamomile, ginger, hibiscus, and chrysanthemum have been recognized for their bioactive complement of aldehydes, ketones, esters, acids, and pyrazine compounds (Xie et al., 2023) and used as ingredients in functional beverage production (Maleš et al., 2022). While sports drinks mostly contain electrolytes and carbohydrates, energy drinks contain a much larger number of constituents among which an important place is taken by caffeine, catechin, amino acids, and vitamins, followed by different plant sources for herbal substances such as ginseng, ginkgo, guarana, and yerba mate (Maleš et al., 2022)

Emulsifiers in beverages are employed to bring aqueous and oil phases together improving their texture and appearance and flavors. Flavored fruit and milk drinks as well as soups and sauces often contain oil-based ingredients or flavors that may separate on storage. These need to be kept emulsified and suspended. Plant-based milk products too face this challenge. Lecithin, monodiglycerides, and gums like guar or carrageenan obtained from natural sources are widely used in such applications and have nutritional value making them good examples of nutraceutical additives (List, 2015, Chemsino, 2024, 2025). However, emulsifiers like polysorbates used in fruit drinks are less easily replaced. Although hydrocolloids like gum arabic and xanthan can act both as emulsifiers and stabilizers, they also increase viscosity and interfere with clarity that may be disliked by consumers (Gossinger et al., 2018).

12.2.2 Bakery and confectionery

Both the baking and confectionery industry use a number of food additives like emulsifiers, preservatives, sweeteners, colorants, flavor enhancers, thickeners, stabilizers, enzymes and leavening agents. Additives in baking improve quality, texture, flavor, appearance, nutritional value besides extending shelf life and assist during processing.

Emulsifiers help in bringing together the basic ingredients of baking—flour, water, fat/oils, leavening agents, etc.—and disperse them uniformly throughout the dough and enhance the aeration process resulting in lighter, fluffier textures in cakes and bread. They also smoothen textures, reduce viscosity, inhibit "bloom" by controlling fat crystallization, reduce stickiness by lubrication, and soften the chewing gum base, thus playing important part in confectionery industry. Lecithin, a natural by-product of oil refining industry, is the most commonly used emulsifier in baking and confectionery industries. Both baking and confectionery use many synthetic emulsifiers like monoglycerides, polysorbates, sorbitan esters, diacetyl tartaric acid esters of monodiglycerides (DATEM), polyglycerol polyricinoleate (PGPR), and sodium stearoyl lactate (SSL). They are often derived from natural fats or other compounds and perform essential functions like crumb softening (monodiglycerides, SSL), improve structure of yeast raised products (SSL), dough strengthening (SSL, DATEM), and overcome adverse flavor, color, and stickiness when higher lecithin concentrations are used. Liquid emulsifiers aid in processing of candies by smooth release when sprayed on molds and machinery. One emulsifier alone usually does not posses all these characteristics and often combinations need to be used (Stampfi and Nersten, 1995). Among nutraceutical alternatives to emulsifiers in confectionery, gum arabic has been used to coat confectionery that can enhance flavor, reduce stickiness, and provide a sheen. By increasing gum arabic to high levels (46%–85%), soft or hard candies with very low sugar content can be produced (Alland & Robert, 2017). Gum arabic is quite versatile, functioning as texturizer, emulsifier, anticrystallization agent, and stabilizer (Rajni, 2023). Pectin has been used in products like jellies, gummies, and fruit fillings to create desirable textures, emulsify, improve shelf life, and enhance flavor release (Reichembach and Petkowicz, 2021).

Preservatives like calcium propionate and potassium sorbate, benzoates, and acetic acid inhibit mold and yeast growth in bread and cakes. Essential oils like thyme, oregano, and cinnamon can substitute some of the synthetic preservatives, besides adding their unique aroma to the products.

Traditional sweeteners in baking cover a wide range: granulated and brown sugar, fructose, liquid sweeteners like honey, maple syrup and corn syrup. Nonnutritive an low-calorie sweeteners like stevia, erythritol, sucralose, saccharin, neotame, xylitol, cyclamate, aspartame, and acesulfame K are also approved for use. Most of these are much more sweeter than sugar and are suitable for baking (Mariotti and Alamprese, 2012, Bakerpedia 2025), but warnings have been sounded about heating foods containing sucralose which might produce potential toxic compounds (Eisenreich et al., 2020). Neotamme, 7000–13,000x sweeter than sucrose, has been specially developed for baking and does not leave metallic or bitter aftertaste, unlike scaccharin or acesulfame K. Stevia is often used with erythritol in baking, but is not useful in meringues. A common critique of these class of sweeteners for baking is that they do not give the body and texture of products made with sugar (Bakerpedia, 2025). Monellin is not useful for baking as it decomposes above 50°C, but thaumatin being heat stable has been reported to be used for reducing the amount of sugar in cakes and cookies (HHOYA, 2025).

Natural or organic colorants in baking and confectionery have the same advantages and disadvantages discussed in Chapter 8. Synthetic colorants like Allura red, Erythrosine, Ponceau4R (Red), Tartrazine, sunset Yellow, quinoline yellow (yellow), Brilliant blue, Indigo carmine, Fast green (Blue and green) are commonly used in baking and confectionery, more so in the latter. Natural colorants like Annato, Curcumin (yellow), beetroot red, Carmine, Paprika (red and orange), Anthocyanins (Red, purple, and blue), chlorophyll (Green) have replaced synthetic colorants in many applications, but are limited by cost and lower vibrancy. Flavor enhancers too are commonly used in baking, which thankfully are mostly from natural sources. Monosodium glutamate (MSG), disodium 5'-ribonucleotides, and yeast extracts are used to boost umami flavors. The 5'-ribonucleotides are obtained by fermentation using *Corynebacterium* spp. (Elhariry et al., 2004), *Bacillus subtilis*, other bacteria, and fungi (Ouyang et al., 2025), while brewer's or baker's yeast is the source of yeast extract. Plant-based natural functional enhancers are also regularly used in baking like malt extract, spices, herbs, and hydrolyzed vegetable proteins. Acidulants like citric, malic, and lactic acid have long been used in gummies, sour candies, fruit flavored and dairy flavored candies, and caramels (Berry, 2001), but citric and malic acids are mostly synthesized rather than sourced from nature.

Thickeners and stabilizers improve texture, consistency, and shelf-life of baking and confectionery products by increasing viscosity, moisture retention, and hydration. They also form gels that enhance mouthfeel and provide structure especially when gluten free flours are used. Starches both in native and modified are used both in baking and confectionery industry (Hadnađev et al., 2018). Leavening agents in baking occupy a central role in providing the characteristic texture, softness, and appearance of bakery products. Most cakes and pastries are raised using baking soda and baking powders because their simple chemical nature can be argued as natural leavening agents. By choosing the acidulant part of the baking powder between phosphates, citrates, tartrates, and other such compounds, it is possible to obtain quick, slow, intermediate, or mixed modes release of carbon dioxide during baking depending on the necessity. However, biological leavening systems that can be held as purely natural leavening agents are also equally popular and are indeed the first choice for bread making. These could be yeasts like *Saccharomyces cerevisiae* or a mix of natural microflora (*Saccharomyces* spp. and *Lactobacillales* found in the sourdough starter or the baker's hands (Reese et al., 2020). Leavening agents are also popular in confectionery to create light, airy texture and structure that can be obtained by incorporating air mechanically or by steaming that by pass the need for synthetic leavening agents. The texture of the popular fermented Indian foods like "Idli" and "Dosa" are due both to biological leavening and the steaming process (Mukherjee et al., 1965).

Bread, being a major source of nutrition is commonly fortified with Vitamins (B_9, C, D_2, D_3), minerals (K, P, Ca, Mg, Fe, Zn, Cr, Se), fiber, proteins, and polyphenolic compounds (Kaim and Goluch, 2023). Many of these vitamins are obtained from natural sources by microbial fermentations or extracted from animal sources (Vandamme, 1992) or plants (McKee and Latner, 2000).

12.2.3 Dairy and dairy alternatives

The modern dairy industry, too, is a prolific user of food additives like stabilizers and thickeners, emulsifiers, preservatives, antioxidants, acidity regulators, sweeteners, colorants, and nutritional fortifiers.

The appearance and texture of many dairy products are often a function of the emulsified state of milk fat, specifically, the stability of the colloidal fat droplets and the ability to lower interfacial tension between the aqueous and fat phases. Emulsifiers play an important role in this and low molecular weight emulsifiers like mono- and diglycerides are mostly used for the purpose (Lal et al., 2006). Stabilizers and thickeners are responsible for the creamy texture of yogurts, puddings, and milk shakes or milk-based drinks and also help in keeping the emulsion stable. Dairy industry mostly uses natural or nutraceutical substances such as Pectins, gelatin, carrageenan, and various gums for stabilization and to thicken many milk-based drinks besides achieving gelation. Both casein and whey proteins can themselves act as emulsifiers besides plant proteins and their emulsifying actions have been recently reviewed by Kim et al. (2020). Emulsification by homogenization is a standard practice in dairy to achieve desirable product characteristics (Köhlera and Schuchmann, 2011), and biopolymers such as proteins and polysaccharides are valued as emulsion stabilizers Yvonne and Jideani (2018)

Natural lactic fermentations by using lactic acid bacteria (LAB) to convert lactose to lactic acid helps in preservation of many dairy products obviating the need for external preservatives. The preservation is both due to lowering of the pH to deter spoilage and pathogens and the production of antimicrobial compounds like bacteriocins. LAB fermentations have indeed been explored as a means of recovering antimicrobial and antioxidant compounds (Khubber et al., 2022). Besides LAB fermentation, chemical preservatives like benzoates, sorbates, propionate are commonly used in dairy products, mostly to control mold and yeast growth (Bhatti, 2019). Natamycin, an antibiotic produced by *Streptomyces natalensis*, is legally approved for use in dairy products by FDA and has a GRAS status. The antibiotic Nisin, as well as other bacteriocins, can be potentially useful in dairy industry. Plant extracts as well as essential oils such as those from rosemary, basil, clove, and oregano, cinnamon, garlic, lemongrass, sage, grape seed, cayenne, and green pepper and many others have been explored as preservatives in dairy products (El-Aziz et al., 2023).

Milk is a natural source of antioxidants of both fat- and water-soluble types. Among the lipophilic antioxidants, conjugated linoleic acid, α-tocopherol, β-carotene, vitamins

A and D3, and coenzyme Q are the major compounds, while the hydrophilic antioxidants are proteins, peptides, vitamins, minerals, and trace elements. A negative correlation between milk consumption and the incidence of diet-dependent diseases confirms that the consumption of milk and dairy products delivers health benefits (Grazyna et al., 2017). Some synthetic antioxidants like BHA and BHT are used in products like anhydrous milk fat, but with such a wealth of natural antioxidants already present in milk, they are being replaced by natural antioxidants from plant sources like rosemary and fermented extracts (Siwach et al 2016, Gad and Sayd 2015).

Acidity regulators in dairy are often produced during fermentation processes for yogurt and cheese making. Besides these, citric and lactic acids along with their salts as well as phosphates of sodium and potassium have been used with $CaCO_3$ carbonate and $MgCO_3$ to control pH, improve stability, desirable flavor profile, texture, prevent microbial growth, and modify milk proteins. Citric acid also is used as an antioxidant and aroma compound in dairy products but replacing it with lactic acid (from LAB) can alter sensory characteristics and consistency in production (Soni et al., 2024).

Sweeteners are extensively used in dairy industry. Sucrose, fructose, and xylitol are used to provide bulk and volume to dairy products besides the obvious sweetening property. In addition, sweeteners that add little calories but provide intense sweetness such as aspartame, sucralose, acesulfame K, and stevia have also been used. Sweeteners are used in ice cream, flavored milk, and notably in traditional Indian dairy products. Major traditional dairy sweets in the Indian market are burfi, sandesh, peda, kheer and their variants, basundi, and gulabjamun. Some of these sweets have been developed in lower calorie versions as a diabetic range of dairy sweets that are being commercially marketed. A number of attempts have been also made to replace bulk sweeteners with intense sweeteners in traditional Indian dairy sweets like Peda, basundi, rabri, etc. in most of which synthetic sweeteners have been dominant, but occasionally stevia has been used (Rao and Pagote, 2018). Negative effects on a number of health parameters have been observed in Mexican children when dairy products with either sucrose, fructose, HFCS, sucralose and acesulfame K sweeteners were consumed (Briones-Avila et al., 2021), but positive effects were seen when stevia, polydextrose or modified starches were used. Texture and fat in the food matrix play an important role in sugar-reduction studies because both highly influence sweet taste perception, clearly indicating the need for more research in this area (McCain et al., 2018).

Fortunately, natural colorants have been replacing synthetic ones in dairy products due to good response from the industry to consumer demands. Natural colors such as annato, carotenoids, chlorophylls, anthocyanins, betalains, and caramels have been employed in a host of dairy products like, butter, cheese, yogurt, cream, fermented milk products, and ice creams (Singh et al., 2023). pH level in dairy products is a crucial determinant in selection of natural colorants and needs to be compatible with the product and application (Imbarex, 2025):

12.2.4 Meat, poultry, and seafood products

Around the world, chickens, pigs, cattle(beef), and fish/seafood are the most common animal species consumed by humans. All animals need to be slaughtered, dressed, and portioned appropriately before being processed and sold for consumption. Fresh meat is the most preferred way of consumption and processing into frozen, canned, or cured/dried products that have longer shelf life is comparatively lesser. Chilling, icing, and refrigeration are the common ways in which flesh foods are preserved in fresh form, which require almost no help from food additives for extending shelf-life. Fish is commonly preserved in ice or slurry ice and addition of antibiotics or preservatives to ice has been shown to extend shelf life. Antibiotics are no longer permitted to be added to ice, but sorbate is being used in ice commercially (Vesta~VAR, 2025). The advent of modified atmosphere packaging (MAP), however, has brought in a paradigm shift in the way fresh flesh foods are sold. MAP is mainly dependent on the customized mix of gases and low temperatures to decrease the growth of microorganisms. Nevertheless, carbon monoxide, antioxidants, organic acids, and biopreservatives are sometimes used as adjuncts in MAP to increase its efficacy and create a "hurdle technology" effect. Most antioxidants for MAP are nutraceuticals like vitamin C, phenols, tocopherols, and extracts of spices and herbs such as oregano, rosemary, sage, thyme, black pepper, etc. (Manessis et al., 2020). Alternatively, plant essential oils have been incorporated into chitosan-based coatings or food packaging material microstructure in a form of "active packaging" (Sharma et al., 2026). Spice and herb marinades for flesh foods not only increase their palatability, but also their shelf-life by their antioxidant and antimicrobial properties (Latoch et al., 2023). However, spice and herb marinades in combination with MAP of flesh foods need caution as such conditions could lead to growth of toxin producing Clostridium botulinum (Virginia Cooperative Extension, 2020)

The other major ways in which meat and fish are processed are curing and smoking, which are traditional methods bordering on the prehistoric. Curing of hams and sausages uses a food additive that has become contentious over the years. Nitrites used extensively in cured meats serve two purposes: suppression of lipid and protein oxidation and, even more importantly, limiting the growth of the pathogenic *Clostridium botulinum.* Nitrosamine production from use of nitrite cured meats is well known and when combined with smoking as is often the case, polycyclic aromatic hydrocarbons (PAHs) from smoke and heterocyclic amines (HCAs) from high-temperature cooking are formed. All of these compounds are either toxic or carcinogenic to humans (Iko Afé et al., 2021, NIH, 2017). Hence, attempts have been made to use plant extracts, many of which contain considerable quantities of nitrate that can be reduced to nitrites by many microbes. Parsley, celery, Swiss chard, blueberry, beetroot powder, grape seed extract, etc. have been explored as nitrate alternatives that are summarized by Shakil et al. (2022). Organic acids, notably lactates have been shown to improve color stability, flavor and shelf life of cured meats (McClure et al., 2011). The flavor and color developed by traditional smoking process are difficult to substitute but the use of liquid smoke rather than high temperature smoking has been in vogue since 1895. Although some liquid smoke products were determined to be genotoxic by in vitro testing, in vivo testing showed that they were not of concern (EFSA, 2007). Phosphates are often added to frozen meat and seafood products to enhance quality, texture, and shelf-life. Although high dietary intake of phosphates can lead to hypocalcemia and phosphate accumulation in kidneys, especially in elderly populations, phosphates are generally recognized as safe when consumed in moderation by healthy individuals (Panseri et al., 2020). Several plant-, fungal-, and algal-based ingredients, notably *Moringa olifera* leaves and *Xylopia aethiopica* fruits, have been evaluated as inorganic phosphate substitutes in meat products with varying degrees of success, based on their antioxidant properties, while citrus fiber and starches have been explored as possible phosphate substitutes for functionality of the latter (Tabanty et al., 2023, Molina et al., 2023)

A class of MFP products that are extremely popular worldwide and make extensive use of additives are the "Coated" products. These are appropriately portioned chicken and seafoods that are dipped in batter and coated with breadcrumbs and that are briefly deep fried, flash frozen, and marketed after packaging. The batter and breader premixes market was valued at nearly 3 billion USD in 2025 and growing at 6.3% CAGR (Mordor Intelligence, 2025). Such products use hydrocolloids, starches (regular and modified), emulsifiers, acidity regulators, fats, dairy, proteins flavor enhancers, anticaking and leavening agents. Antioxidants are also added to overcome rancidity and fibers to reduce oil absorption in the fried product, especially in seafoods. Although promising potential, present use of nutraceuticals in coated products appear to be limited.

Surimi is another processed fish product that makes extensive use of food additives Natural cryoprotectants like polysaccharides, proteins, and polyols are used in surimi manufacture to protect myofibrillar proteins from denaturation and damage caused by ice crystals and water migration during frozen storage.

Common additive examples include sucrose, sorbitol, trehalose, maltodextrin, and polydextrose, to increase osmotic pressure, reducing freezable water, and directly interacting with proteins (Walayat et al., 2022). Omega-3 fatty acids, Antarctic krill oil emulsions, fish bone biocalcium, seaweed extract, and herbal extracts are some of the nutraceuticals added to surimi (Roy et al., 2021).

12.2.5 Sauces, dressings, and condiments

This group of food products uses several additives like preservatives, acidity regulators, antioxidants, emulsifiers, thickeners, flavor enhances, and food colors.

Common preservatives used are sodium benzoate, potassium sorbate, vinegar, salt, and sulfites. Some are used in combination like benzoate and sorbates against, bacteria, yeasts, and molds. Vinegar and lemon juice readily serve as natural preservatives in sauces apart from herbs and spices such as rosemary, garlic, oregano, tamarind. The herbs and spices in addition also serve as antioxidants (Al-Habsi et al., 2025), besides adding their own unique flavors. Traditional Indian gravies use a combination of aromatic spices like turmeric, cumin, coriander, chili powder, and fresh herbs such as ginger, garlic, onions for color and flavor, along with warm spices such as cardamom, cinnamon, and cloves for complex warmth, which too serve as preservatives and antioxidants. The fresh herbs often form the base of such gravies.

Natural emulsifiers like egg yolk, soy lecithin, and mustard are already used in manufacture of sauces and dressings. In addition, gums such as xanthan and guar are also used along with synthetic ones like polysorbates and mono- and diglycerides.

Egg yolk commonly used in making mayonnaise and such sauces has often been sought to be replaced to make a healthier but functional products. Chitosan, baicalin, and gamma irradiation have been explored as possible alternatives in such attempts that have been summarized by Zanjani et al. (2019). Thickeners traditionally used in sauces and dressings are already from natural sources like corn starch, arrowroot, and tapioca starch. Coconut milk is another natural ingredient that finds extensive use in south Indian and coastal cuisine for both sweet and savory preparations. It also confers nutritional and therapeutic benefits (Gengal et al., 2025). Coconut milk, when industrially produced, is preserved by canning, which does not require any additives. Yogurt, especially Greek yogurt, has often been used as a thickener traditional cooking to provide a creamy texture and is sometimes combined with corn starch for better functionality (Silva Leal et al., 2024).

Colorants are commonly used in sauces, dressings, and condiments to meet consumer expectancies, consistency, restore color degradation due to processing, and enhance perceived quality and flavor intensity. While synthetic colors are often used for these applications due to their stability and effectiveness over a wide range of conditions, natural colors are increasingly replacing the former which has been discussed in earlier sections. Processing conditions and ingredient interactions can often alter natural colors when used and need to be used carefully (Sezgin and Ayyıldız, 2017). Synthetic colorants such as azo dyes, indico carmine, and fast green FCF can be substituted with natural options like beta-carotene, caramel, turmeric, and paprika extract (Gemini, 2025).

12.2.6 Snacks and ready-to-eat foods

Snacks and ready-to-eat foods (SNRTE) are a diverse class of foods with origins in all major food groups. They may require cooking by way of a variety of cooking techniques like boiling, baking, frying, microwaving, etc. or sometimes need no cooking (ready to eat) or are ready to use requiring minor preparation like reconstitution or warming etc. They can be sweet, savory, spicy, and many in-between variants. It is only natural that their manufacture requires some form of additives or other. FSSAI (2025) classifies foods into 18 major categories, of which category 15 specifically deals with savory SNRTE. But other categories can also contain SNRTE. However, the additives added to most SNRTE are quite similar, such as preservatives, antioxidants, texture modifiers (emulsifiers, stabilizers and thickeners), sweeteners, colorants, and flavor enhancers. Contrary to general thinking, many SNRTE can also be high moisture and high a_w foods like fruits, vegetables, and dairy products (yogurt, smoothies, cottage cheese). In many cases by combining alternative preservation methods such as UHT sterilization, freezing, canning, or nitrogen filling, many RTE have been marketed as preservative free. Traditional Indian foods that required lengthy cooking techniques are now packed in retort pouches and sterilized or frozen thus extending the traditional definition of Ready to eat foods. While such foods entirely bypass the need for preservtives, other popular SNRTE like extruded snacks, confectionery, fried snacks etc, do use many additives which can be explored for substitution with natural nutraceutical alternatives.

Benzoates, propionates, sulfites, nitrites, and sorbates are the common preservatives along with antioxidants (BHA/BHT, propyl gallate, etc.) found in SNRTE. Organic acids like lime juice, vinegar, and raw mango powder (aam choor) are popular acidulants and the latter has become quite popular commercially in its own right. Additionally herbal powders like moringa leaf, ginger, cinnamon, garlic, onion, etc. are known antimicrobials and have been used in khakhra, bhakri, ginger bites, flavored crackers, popcorn, savory granola bars, etc. (Gemini 2025b). Many of these natural nutraceuticals also are well-known antioxidants and prevent oxidation of SNRTE rich in fats (Lagouri and Boskou 1995, Embuscado, 2015). Emulsifiers obtained from plant and animal sources (lecithins) and thickeners and stabilizers from gums (acacia, carrageenan, guar, xanthan) are commonly added to processed foods such as mayonnaise, ice cream, beverages, jellies, baked goods, etc. (Angioma Alliance, 2020). Sweeteners in SNRTE are ubiquitous and can be seen in the ingredient list of even savory snacks. Sugar is often the common choice, but stevia, monk fruit, date syrup, agave nectar, glycyrrhizin, tagatose, etc. are also being used along with sweet tasting herbs like cinnamon, star anise, fennel, and tarragon (Saraiva et al., 2020). However, recommendations surrounding nonsugar sweetener consumption are inconclusive and conflicting, with scarce evidence on long-term effects, which should make us pause in adopting their extensive use (Hedrick et al., 2023). The substitution of synthetic colorants and flavor enhancers with natural nutraceuticals has already been dealt with in earlier sections.

12.3 Cross-industry additives with nutraceutical potential

The use of food additives in various types of food industries shows that additives have similar functions across the major food types and it is rare to find food additives that are unique to only one type of food industry. It is an instance of function following form and as long as a particular function is required in a prepared food, the requirement can be met by selecting the correct food additive. Thus, many nutraceuticals find applications across different food industries. Table 12.1 readily illustrates this fact.

12.4 Industrial considerations for switching to nutraceutical alternatives

Food industry already has innumerable food additives for nearly every conceivable purpose(s). Hence, the question arises on the need to switch to nutraceutical alternatives. The arguments in favor of nutraceutical additives in foods are well-known: added dimension of benefits of health and nutrition, consumer demand for clean label products, and a back-to-nature movement. But how does this benefit the food industry in terms of equivalent or superior performance, economic or financial advantage, ease of processing and handling, etc. needs to be examined. As the numbers and purposes of food additives are huge, broad generalizations in favor of or against nutraceutical alternatives would be unwise.

Table 12.1 Nutraceutical alternatives to food additives and industrial applications.

Class of food additive and sub class	Nutraceutical alternative	Food industries where it can be used, with reference
Preservatives		
	Ascorbic acid (*Citrus limon*)	Beverages, baked goods, cured meats, jams and jellies, and canned goods, Muthu Lakshmi et al. (2025)
	Citric acid (*Citrus limon*), vinegar, lactic acid (bacterial fermentation)	Beverages, baked goods, dairy products cured meats, jams and jellies, and canned goods
	Herbs: Rosemary (*Salvia rosmarinus*), Thyme (*Thymus vulgaris*) Cloves (*Syzygium aromaticum*) Garlic (*Allium sativum*) Oregano (*Origanum vulgare*)	Meat, dairy, baked goods, sauces, and seafood products, Veenstra & Johnson (2019)
	Natural oils (olive oil, coconut oil, mustard oil)	Seafood, poultry, and meat products, fried foods, crackers, desserts, candies, and nondairy creamers, pickles, Martinez-Zamore et al. (2021)
	Salt and sugar	Cured foods, jams, and jellies
	Chitosan	Fruit and vegetable, meat, dairy, and beverage industries, Duan et al. (2019)
Antioxidants		
	Tocopherols (most edible oil seeds)	Dairy products, cereals and baked goods, meat products, confectionery, oils and fats, salads, and dressings, Lourenço et al. (2019)
	Ascorbic acid, ascorbyl palmitate,	Baked goods, beverages, confections (candies), dairy products, frozen foods, jams and jellies, meats (especially cured meats and fresh ground meat), fish and seafood, fruits, and vegetables, Cort (1982)

(*Continued*)

Table 12.1 (Continued)

Class of food additive and sub class	**Nutraceutical alternative**	**Food industries where it can be used, with reference**
Spice and herb extracts	Oregano (*Origanum vulgare*), rosemary (*Salvia rosmarinus*), thyme (*Thymus vulgaris*), cloves (*Syzygium aromaticum*), garlic (*Allium sativum*), basil (*Ocimum basilicum*)	Bakery, dairy, meat, fish, oil, confectionery, Kapadia et al. (2022); Mashabela et al. (2022); Al-Habsi *et al.* (2025)
Phenolic compounds	Gallic acid (*Caesalpinia spinosa*) Oak galls (*Quercus* spp.) Ferulic acid (brewers spent grain and wheat bran) Chlorogenic acid (green coffee beans, apple pomace, and coffee grounds)	Beverage, meat, dairy, baking, and fruit/vegetable processing industries, Kapadia et al. (2022)
Flavonoids	Quercetin (Quince *Cydonia oblonga Miller*) Catechin (green tea leaves and green tea waste)	Functional foods, dietary supplements, and beverage additives, Pool et al. (2012); Rodríguez et al. (2021)
Carotenoids	Carrots Black carrot *Daucus carota ssp. sativus var.* atrorubens Alef,	Dairy, processed meats, confectionery, baked goods, beverages and juices, fortified foods, Zakynthinos and Varzakas (2016)
Essential oils	Oregano, thyme, clove, cinnamon, rosemary, citrus oils (lemon and orange) ginger (*Zingiber officinale*)	Dairy, meat, bakery, seafood, confectionery, fruit/vegetable industries, Adetunji and Sharifi-Rad (2023)
Citric acid (as chelator of metal prooxidants)	Citrus fruits (*Citrus limon*)	Soft drinks, juices, jams, jellies, processed meats, dairy products, baked goods, canned goods, Allen (2015); Swain et al. (2012)
Curcuminoids	Turmeric (*Curcuma longa)*	El-Saadony et al. (2023)
Tanins	Oak bark (*Quercus* spp.) Mimosa (*Acacia measnsii*), Chestnut (*Castanea sativa* Mill), grape seeds, and skins (*Vitisvinifera*)	Control astringency, bitter flavors, clarity in wines, teas, juices, functional foods, Singh and Kumar (2019); Cosme et al. (2025)

(*Continued*)

Table 12.1 (Continued)

Class of food additive and sub class	Nutraceutical alternative	Food industries where it can be used, with reference
Emulsifiers, stabilizers, and thickeners		
	Lecithins egg yolk and soyabean oil (*Glycine max*)	Bakery products, chocolates, dairy, confections, fat-based foods, List (2015)
	Mustard (*Brassica juncea*)	flavoring meats, sauces, and dressings, preservative in processed foods, thickening agent in processed meat and sauces, (Wu et al., 2015, Poyil et al., 2022)
	Protein (eggs, soya)	Dairy, meat, bakery, beverage, confectionery industries, Liu et al. (2024a,b); Kim et al. (2020)
	Caseins and whey proteins	Dairy, meat, bakery, beverage, confectionery industries, Kim et al. (2020); Zhu (2024)
	Carrageenan Red seaweeds (Rhodophyta)	Dairy, meat, bakery industry, (Udo et al., 2023)
	Guar gum (*Cyamopsis tetragonoloba*)	Dairy, bakery, meat, beverage industries, Mudgil et al. (2014)
	Xanthan gum (*Xanthomonas campestris*)	Dairy products, dressings, bakery items, low-calorie foods, beverages, sauces, (Chaturvedi et al., 2021)
	Agar-agar (*Gelidium* spp.)	Desserts, confectionery, dairy, beverages, canned food, jams and preserves, baking, Pandya et al. (2022)
	Pectin (citrus peels)	Jams and jellies, dairy, confectionery, low calorie foods, bakery products, edible films and coatingd, pre-biotics, Sharma et al. (2026)
	Gum acacia (gum arabic) (*Acacia auriculiformis*)	Confectionery, beverages, baked goods, Rajni (2023)

(*Continued*)

Table 12.1 (Continued)

Class of food additive and sub class	Nutraceutical alternative	Food industries where it can be used, with reference
Colorants		
Yellow, yellow-orange, yellow-brown	Turmeric (*curcuma longa)*, Marigold (*Tagetes patula)*, Weld (*Reseda luteola L*), Carrots (*Daucus carota subsp. sativus*), Passion fruit (*Passiflora edulis f.flavicarpa*), Annatto *(Bixao orellana)*, Riboflavin	Dairy, baked goods, confectionery, beverages, pasta, functional foods, processed fish products, Kalsec (2025); Givaudan (2025)
Orange	Paprika (*Capsicum* spp.) Annatto (*Bixao orellana*), saffron (*Crocus sativus*)	Dairy, baked goods, confectionery processed meats, cheese, sauces and dips, snacks, spice mixtures, ready-to-eat meals, Novais et al. (2022)
Red, red-brown Betalains: betaxanthins: yellow orange Betacyanins (betanin): red-violet	Carminic acid (Cochineal) (*Dactylopius coccus*) Beetroot (*Beta vulgaris*) Spinach (*Spinacia oleracea*) Amaranthus (*Amaranthus tricolor, A. cruentus*) Prickly pear (*Opuntia* spp.) Dragon fruit (*Selenicereus undatus*)	Dairy, confectionery, beverages, bakery, snacks, Pasta, cereals meat products, Novais et al. (2022); Azeredo (2009); Adetunji and Sharifi-Rad, 2023; Nabi et al., 2023
Red Flavonoids, anthocyanidins, anthocyanins	Strawberries (*Fragaria ananassa*) Rraspberries (*Rubus idaeus*), Cranberries (*Vaccinium subg. oxycoccus*), Cherries (*Prunus avium* L.)	Jams, jellied, soft drinks, dairy products, confectionery, bakery goods, Khoo et al. (2017)
Blue	Butterfly pea (*Clitoria ternatea*) Grapes (*Vitis vinifera*). Blueberries (*Vaccinium sect. Cyanococcus*), Black currants (*Ribes nigrum*) Red cabbage Brassica oleracea var. capitata f. rubra Spirulina (*Arthrospira platensis*)	Herbal teas, cocktails, confectionery, frozen desserts and drinks, Jams, juices, nutraceuticals beverages, confectionery, dairy products, fruit preparations, meat and fish processing, food supplements, Neves et al. (2021)

(*Continued*)

Table 12.1 (Continued)

Class of food additive and sub class	Nutraceutical alternative	Food industries where it can be used, with reference
Purple	Anthocyanidin Black carrot *Daucus carota ssp. sativus var. atrorubens* Alef, Sweet potato *Ipomoea batatas* (L.) Lam	Confectionery, dairy, beverages, bakery products, Novais et al. (2022); Plant lipids (2025)
Green	Chlorophyll a and b Cyanobacteria (*Cyanobacteriota* spp.) Spinach (*Spinacia oleracea*) Matcha (Tea *Camellia sinensis* (L.) Kuntze)	Beverage, dairy, confectionery, bakery industries, Plant lipids (2025); Sun et al. (2023)
Brown	Caramel (sugar caramelized to different extents) Black tea (*Camelia sinensis*)-based solutions	Beverages, baked goods, confectionery, dairy products, savory foods, pet foods cereals, (Sengar and Sharma (2014)
Black	Caramel Coconut shell carbon Squid ink (*Loligo* spp.) Black sesame seeds (*Sesamum indicum*) Black cocoa powder (*Theobroma cacao*)	Baking, confectionery, beverage, sauce, snacks industries, Corecarbons 2023; Dossou et al. (2022); Biehl and Ziegleder (2003)
Sweeteners		
	Cane sugar (Saccharum officianarum)	Provide sweetness, reduce calories, and improve various food properties, including texture, moisture retention, and browning, beverages, bakery products, dairy, sweets, Confectionery, and chewing gum, Sylvetsky and Rother (2016)
	Stevia (*Stevia rebaudiana* Bertoni)	Zero-calorie sweetener beverages, dairy products, bakery goods, and confectionery, Naik and Poyil (2022)
	Monk fruit *Siraitia grosvenorii* (Swingle) C	Zero-calorie sweetener beverages, baked goods, desserts, dairy, and savory items, Chen et al. (2024)

(*Continued*)

Table 12.1 (Continued)

Class of food additive and sub class	Nutraceutical alternative	Food industries where it can be used, with reference
	Katemfe fruit (*Thaumatococcus daniellii* (Benth)	Zero-calorie sweetener, flavor enhancer and modifier, soft drinks, health bars, dairy alternatives mask bitter taste, and balance sour notes. Baked goods and supplements, Dalavi et al. (2024)
	Yacon roots (*Smallanthus sonchifolius*)	Low-calorie sweetener, fat replacer, and prebiotic ingredient in yogurts, diet cookies, oatmeal, and pancakes, Yan et al. (2019)
	Honey	Beverages, baked goods, confectionery, spreads, like yogurts and honey- containing flakes, Bellik and Iguer-ouada (2013)
	Maple syup (*Acer saccharum*)	Baked goods, pancakes and waffles, desserts, glazes, and some savory dishes
	Date syrup (*Phoenix dactylifera)*	
Flavorings		
	There is a huge variety of flavorings used in food as additives. Most of them are obtained from natural sources by techniques of distillation or extraction.	Reshna et al. (2022); Singh and Sudha (2024)

As nutraceutical additives confer advantages beyond the ad hoc "food additive" purpose, in terms of extra nutrition or health benefits, the processor will be able to increase nutrition and claim health benefits for the product. For example, vegetable lecithins when substituted for mono- or diglyceride emulsifiers confer added caloric nutrition; the use of citrus fruits as acidulant or antioxidant also lowers oxidative stress and inflammation due to its ascorbic acid content (Vignesh et al., 2024). Nutraceuticals can also enhance shelf life, improve physical properties, and create value addition. Herbal preservative additives like rosemary, garlic, and oregano apart from their primary purpose of preservation, and flavoring of food, provide strong antioxidant, anti-inflammatory, mood enhancing and immune boosting properties that support heart and immune health (Iommelli et al., 2025). Thus, they add an extra dimension of functionality to the processed food enabling marketing of the product on multiple levels (Aurea biolabs, 2025).

Many food processing by-products are rich in nutraceutical compounds like proteins, fibers, vitamins, and antioxidants. By recovery and use of such nutraceuticals by-products, new, value-added products are obtained improving resource utilization and creating additional value streams (Ran et al., 2019, Mitraka et al., 2021) to the processor. They also lower the cost of waste disposal by decreasing the amount of organic waste to be disposed or recycled. For example, extending shelf-life a key function of additives can be achieved using packaging film of chitosan, a by-product of shrimp processing, which is nontoxic, biodegradable, and has antimicrobial and antifungal properties (Flórez et al., 2022). Nutraceutical preservatives have also been incorporated into many edible film coatings for fruits and vegetables (Venkatesan and Muniyan, 2024).

Many of the nutraceuticals are biopolymer molecules, like pectins, blackberry polysaccharide, sesame hull polysaccharide, etc., which can allow them to perform multiple functions as emulsifiers, stabilizers, or fat replacers to enhance texture, viscosity, emulsion stability, water retention, antioxidant, and overall product yield and shelf-life (Wang et al., 2024). Dough handling, for example, can be improved by addition of vitamin C or whey proteins that speeds up gluten development, gluten structure, and elasticity, while dough strengthening and stability can be improved by pectin and beta-glucans (Jia et al., 2022, Scappaticci et al., 2024, Fluerasu et al., 2025).

While many of the nutraceuticals may be costlier compared to their conventional additives, using them gives the processor the unique "clean label" advantage, which may off-set the costs of regulatory compliance and labeling requirements. However, conventional and synthetic additives are usually more cost effective than equivalent nutraceuticals. Potassium sorbate, a common antifungal additive, costs Rs. 0.3/g, while nutraceutical antifungals like cinnamon (Rs. 1.6/g) or garlic oil (Rs. 4/g) are much more expensive (September 2025 prices). As most nutraceuticals are biological and agrihorti produce, their supply and prices tend to be volatile and not easily scaled up or down. There is also the question of varying potency of nutraceuticals because of their bio-origin and thus their effective dosage in foods. The added flavor of many nutraceutical additives like ginger/turmeric/ cinnamon can be a bonus in many foods but can also become a hindrance in their wider application.

12.5 Sustainability and market trends

The global nutraceuticals market is a sun rise sector driven by rising consumer health consciousness, increasing demand for preventive healthcare and prevalence of lifestyle diseases. The preventive healthcare aspect is further driven by the rising hospitalization and insurance costs. Increasingly people are customizing their personal nutrition, while innovations in functional foods and increasing product accessibility through e-marketing are boosting consumer choice and awareness. The global nutraceutical market is expected to grow at 10% CAGR till 2032 with the Asia-pacific dominating the market with about 40% share. The increasing popularity of functional foods and beverages within several countries of Asia Pacific has supported the regional market growth. (Fortune business insights, 2025). The lack of defined quality standards for nutraceuticals and the varying regulatory regimes in different countries for nutraceuticals can be a hindrance to further growth of this sector.

As the nutraceuticals are poised for rapid growth in the coming years, a question arises if their present supply levels will be sufficient to meet their projected demand. Perhaps lulled by the ever-increasing demand for nutraceuticals and favorable market size, most surveys have not concentrated on this aspect. The supplements and functional foods subsector of nutraceuticals too is not an exception and most estimates are in terms of value rather than volume. The highly varied nature of nutraceuticals both in the foods and pharmaceutical sector also makes this type of estimations and surveys difficult. Collecting demand and supply data in terms of individual nutraceuticals, say, Oat β-glucans or omega-3 fish oils, and collating them would be needed. Even then, such a collation would not be meaningful due to the diversity in quantity and unit values. At present, it would be safe to assume that the supply of nutraceuticals will not be limiting in the near future. However, the biological variation in the active ingredient(s) of the nutraceuticals w.r.t. cultivars, climate, season, and agroclimatic regions needs to be given due consideration as well as their shelf-life (Ullrich, 2025).

An interesting trend has recently surfaced where Synthetic Active Nutraceutical Ingredients (SANI) have begun to be manufactured. This trend needs further critical examination. Do the nutraceuticals owe their beneficial effects to only one or some of the components in the natural source or to the entire nutraceutical source as a whole? Do the components in the source interact to bring about the observed effect(s) or, are their effects are independent of each other? The possible "matrix effect" needs a deeper understanding before the familiar reductionist approach takes over and nutraceuticals begin to be *manufactured* rather than *grown* (Fig. 12.1).

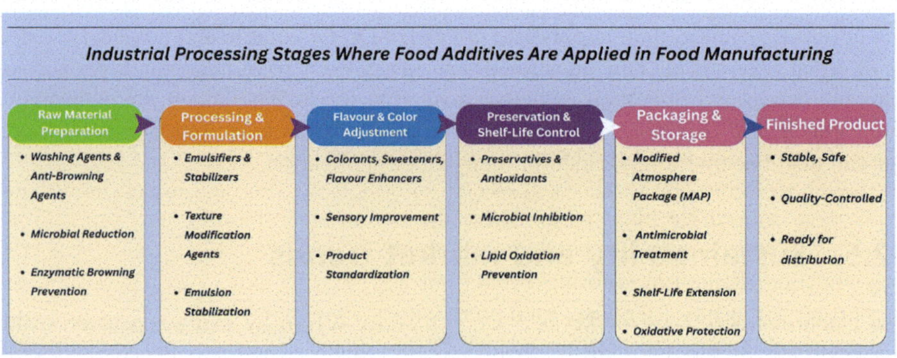

Figure 12.1 Stages of food processing showing where additives are applied, including raw material preparation, formulation, sensory adjustment, preservation, packaging, and final product stabilization.

12.6 Summary

Food additives are used across all sectors of food industry for numerous purposes. In many cases, natural nutraceuticals can be and are being used in place of the synthetic additives. This gives nutraceuticals the twin benefits of the needed additive function and the bonus of nutrition in a "clean label" package. In some cases, nutraceuticals may even out-perform their synthetic counterparts. Nutraceutical food additives are usually obtained/isolated/extracted from natural sources, and hence subject to variations in potency, supply, and costs. Nutraceuticals also need to be competitive in terms of both performance and costs as their synthetic counterparts. Some key questions arise using nutraceuticals as food additives: Can an organic compound, synthesized or obtained by fermentation with GMOs rather than used in its natural form, still be considered as a "nutraceutical"? The natural source may indeed have additional substances in it that confer its observed additive and/or nutraceutical functions. In view of the difficulty in defining "nutraceuticals" by multiple regulatory agencies, standardization is a timely necessity.

References

Adetunji, C.O., Sharifi-Rad, J., 2023. Applications of essential oils in the food industry applications of essential oils in the food industry. Elsevier, Nigeria, pp. 1–359. Available at https://www.sciencedirect.com/book/9780323983402, https://doi.org/10.1016/C2021-0-00817-8.

Alland & Robert (2017). Arabic gum for confectionery coating. Alland & Robert. https://www.allandetrobert.com/info/arabic-gum-for-confectionery-coating/

Al-Habsi, N., Al-Khalili, M., Haque, S.A., Al Akhzami, N., Gonzalez-Gonzalez, C.R., Al Harthi, S., et al., 2025. Herbs and spices as functional food ingredients: A comprehensive review of their therapeutic properties, antioxidant and antimicrobial activities, and applications in food preservation. Journal of Functional Foods 129. Available at https://www.sciencedirect.com/science/journal/17564646, https://doi.org/10.1016/j.jff.2025.106882.

Allen, K.E., 2015. Metal chelators as antioxidants for food preservation. In: Shahidi, F. (Ed.), Handbook of Antioxidants for Food Preservation. Woodhead Publishing, pp. 79–104 ISBN:978-1-78242-089-7.

Angioma Alliance, 2020. Emulsifer list final. Available at https://www.alliancetocure.org/wp-content/uploads/2020/08/EmulsifierList.pdf.

Aurea biolabs. (2025). Shortening the supply chain: Integrating natural food colors into nutraceuticals for smarter, cleaner solutions. shortening the supply chain: Integrating natural food colors into nutraceuticals for smarter, cleaner solutions. Aurea Web.

Azeredo, H.M.C., 2009. Betalains: properties, sources, applications, and stability—A review. International Journal of Food Science and Technology 44, 2365–2376. Available at https://doi.org/10.1111/j.1365-2621.2007.01668.x.

Bakerpedia, 2025. Sugar-free baking. BAKERpedia. Available at https://bakerpedia.com/processes/sugar-free-baking/.

Bantle, J.P., 2009. Dietary fructose and metabolic syndrome and diabetes. The Journal of Nutrition 139 (6), 1263S–1268S. Available at https://doi.org/10.3945/jn.108.098020.

Bassoli, A., Merlini, L., 2003. SWEETENERS | Intensive. In: Caballero, B. (Ed.), Encyclopedia of food sciences and nutrition (Second Edition). Academic Press, pp. 5688–5695. ISBN9780122270550. Available at https://doi.org/10.1016/B0-12-227055-X/01172-X.

Bellik, Y., Iguer-ouada, M., 2013. Honey in the food industry. In: Boukraa, L. (Ed.), Honey in traditional and modern medicine. CRC Press. Taylor and Francis Group, pp. 409–434. Available at https://www.researchgate.net/publication/275966744_Honey_in_the_Food_Industry.

Berry, S.K., 2001. Role of acidulants in food industry. Journal of Food Science and Technology 38, 93–104. Available at https://www.scirp.org/reference/referencespapers?referenceid=2479088.

Bhatti, B., 2019. Use of Preservatives in Dairy Industry. Available at https://www.dssimage.com/blog/use-of-preservatives-in-dairy-industry/.

Biehl, B., Ziegleder, G., 2003. COCOA | production, products, and use. Encyclopedia of Food Sciences and Nutrition (Second Edition) 1448–1463. Available at https://www.researchgate.net/publication/288154164_COCOA_Production_Products_and_Use.

Bocker, R., Silva, E.K., 2022. Innovative technologies for manufacturing plant-based non-dairy alternative milk and their impact on nutritional, sensory and safety aspects. Future Foods 5, 100098. Available at https://doi.org/10.1016/j.fufo.2021.100098.

Briones-Avila, L.S., Moranchel-Hernández, M.A., Moreno-Riolobos, D., Silva Pereira, T.S., Ortega, R.A.E., López, V.K., et al., 2021. Analysis of caloric and noncaloric sweeteners present in dairy products aimed at the school market and their possible effects on health. Nutrients 13 (9), 2994. Available at https://doi.org/10.3390/nu13092994 27.

CAC (Codex Alimentarius Commission) (2024). Joint FAO/WHO Food Standards Programme CAC47/CRD02. 47th session. Part 1. New work proposal on the development of standards for health supplements/ dietary supplements/functional foods and nutraceuticals (9 pp).

Chaturvedi, S., Kulshrestha, S., Bhardwaj, K., Jangir, R., 2021. A review on properties and applications of xanthan gum. In: Vaishnav, A., Choudhary, D.K. (Eds.), Microbial Polymers. Springer, Singapore. Available at https://doi.org/10.1007/978-981-16-0045-6_4.

Chemsino, 2024. Emulsifiers in beverage production: Everything you need to know. Chemsino. Available at https://www.cnchemsino.com/blog/emulsifiers-in-beverage-production-everything-you-need-to-know.html accessed 30.08.2025.

Chemsino, 2025. The role of emulsifiers in instant sauce mixes. Chemsino. Available at https://www.cnchemsino.com/blog/the-role-of-emulsifiers-in-instant-sauce-mixes.html.

Chen, N., Cao, W., Yuan, Y., Wang, Y., Zhang, X., Chen, Y., et al., 2024. Recent advancements in mogrosides: A review on biological activities, synthetic biology, and applications in the food industry. Food chemistry 449, 139277. Available at https://doi.org/10.1016/j.foodchem 2024.139277.

Corecarbons, 2023. Food & Beverages Core Carbons Pvt. Ltd. Available at https://corecarbons.com/food-beverage.html#:~:text=Coconut%20shell%20based%20activated%20carbon,and%20beverages%20industry%20such%20as.

Cort, W.M. (1982). Antioxidant properties of ascorbic acid in foods. In: Seib, P.A., Tolbert, B. M., Ascorbic Acid: Chemistry, Metabolism, and Uses. Advances in Chemistry. 200(22), (pp 533-550) Available at https://doi.org/10.1021/ba-1982-0200.ch022.

Cosme, F.A., Pinto, T., Oliveira, I., 2025. A Comprehensive review of bioactive tannins in foods and beverages: Functional Properties, health benefits, and sensory qualities. Molecules 30 (4), 800. Available at https://doi.org/10.3390/molecules30040800.

Dalavi, M., Handge, S., Pate, V., Raut, H., Kakad, A., Shaikh, M., 2024. Fundamental overview and applicability of thaumatin [Internet]. Int J Pharm Chem Anal 2024. Available at https://doi.org/10.18231/j.ijpca.2024.021.

Dossou, S.S.K., Luo, Z., Wang, Z., Zhou, W., Zhou, R., Zhang, Y., et al., 2022. The dark pigment in the sesame (*Sesamum indicum* L.) seed coat: Isolation, characterization, and its potential precursors. Frontiers in Nutrition 28 (9), 858673. Available at https://doi.org/10.3389/fnut.2022.858673.

Duan, C., Meng, X., Meng, J., Khan, M.I.K., Dai, L., Khan, A., et al., 2019. Chitosan as a preservative for fruits and vegetables: A review on chemistry and antimicrobial properties. Journal of Bioresources and Bioproducts 4 (1), 11–21. Available at https://doi.org/10.21967/jbb.v4i1.189.

EFSA, 2007. (European Food Safety Authority). Risk assessment of the Scientific Panel on food additives, flavourings, processing aids and materials in contact with food (AFC) on the Smoke Flavouring Primary Product—FFB. Available at https://doi.org/10.2903/j.efsa.2007.20r.

EFSA, 2021. (European Food Safety Authority) Panel on Food Additives and Flavourings (FAF). Re-evaluation of thaumatin (E 957) as food additive. EFSA J 19 (11), e06884. Available at https://doi.org/10.2903/j.efsa.2021.6884.

Eisenreich, A., Gürtler, R., Schäfer, B., 2020. Heating of food containing sucralose might result in the generation of potentially toxic chlorinated compounds. Food chemistry 321, 126700. Available at https://doi.org/10.1016/j.foodchem.2020.126700.

El-Aziz, M.A., Salama, H.H., Sayed, R.S., 2023. Plant extracts and essential oils in the dairy industry: A review. Foods and raw materials 11 (2), 321–337. Available at https://doi.org/10.21603/2308-4057-2023-2-579, http://jfrm.ru/en, https://elibrary.ru/REUHRY.

El-Saadony, M.T., Yang, T., Korma, S.A., Sitohy, M., Abd El-Mageed, T.A., Selim, S., et al., 2023. Impacts of turmeric and its principal bioactive curcumin on human health: Pharmaceutical, medicinal, and food applications: A comprehensive review. Frontiers in Nutrition 10 (9), 1040259. Available at https://doi.org/10.3389/fnut.2022.1040259.

Elhariry, H., Kawasaki, H., Auling, G., 2004. Recent advances in microbial production of flavour enhancers for the food industry. Recent Research Developments in Microbiology 8, 15–39.

Embuscado, M.E., 2015. Spices and herbs: Natural sources of antioxidants—A mini review. Journal of Functional Foods 18 (Part B), 811–819. Available at https://doi.org/10.1016/j.jff.2015.03.005.

Fluerasu (Bălțatu), D., Neagu, C., Dossa, S., Negrea, M., Jianu, C., Berbecea, A., et al., 2025. The use of whey powder to improve bread quality: A sustainable solution for utilizing dairy by-products. Foods 14 (16), 2911. Available at https://doi.org/10.3390/foods1416291.

Flórez, M., Guerra-Rodríguez, E., Cazón, P., Vázquez, M., 2022. Chitosan for food packaging: Recent advances in active and intelligent films. Food Hydrocolloids 124 (Part B), 107328. Available at https://doi.org/10.1016/j.foodhyd.2021.107328.

FMI (Future marker Insights inc). (2025). Available at https://www.futuremarketinsights.com/reports/beverage-acidulants-market.

Fortune Business insights. (2025). Nutraceuticals market size, share & industry analysis, by product type (dietary supplements, functional foods, and functional beverages), distribution channel (hypermarkets/supermarkets, convenience stores, online retail, and others) and regional forecast, 2025-2032. Nutraceuticals market size, trends & growth report [2032].

FSSAI, 2025. Food safety compliance system FSSAI. Available at https://foscos.fssai.gov.in/standard-product.

Gad, A.S., Sayd, A.F., 2015. Antioxidant Properties of rosemary and its potential uses as natural antioxidant in dairy products—A review. Food and Nutrition Sciences 6 (1), 179–193. Available at https://doi.org/10.4236/fns.2015.61019.

Gemini. (2025). Gemini-2.5-flash Google [Generative AI] 26.09.2025 AI overview; commercial ready to eat snacks that contain ginger OR garlic OR Rosemary extracts.

Gemini. (2025b). Gemini-2.5-flash Google [Generative AI] 2025.09.29 AI overview: Colorants in sauces, dressings and condiments.

Gengal, G., Zaini, N.S.M., Hussin, S.N., 2025. Nutritional and therapeutic benefits of coconut milk and its potential as a plant-based functional yogurt alternative: A review. Food Science and Human Wellness 14, 9250004. Available at https://doi.org/10.26599/FSHW.2024.9250004.

Givaudan, 2025. Natural Alternatives to Synthetic Yellow. Givaudan Sense colour. Available at https://learn.ddwcolor.com/natural-alternatives-to-synthetic-yellow/#:~:text=The%20most%20versatile%20source%20of,and%20cookies%20or%20cream%20fillings.

Gossinger, M., Buchmayer, S., Greil, A., Griesbacher, S., 2018. Effect of xanthan gum on typicity and flavour intensity of cloudy apple juice. J Food Proc.Preserv 42 (3), e13737. Available at https://doi.org/10.1111/jfpp.13737.

Grazyna, C., Hanna, C., Adam, A., Magdalena, B.M., 2017. Natural antioxidants in milk and dairy products. Int.J.Dairy Tech. 70 (2), 165–178. Available at https://doi.org/10.1111/1471-0307.12359.

Hadnađev, M., Dapčević-Hadnađev, T., & Dokić, L. (2018). Functionality of starch derivatives in bakery and confectionery products. In A. M. Grumezescu & A. M. Holban (Eds.), Biopolymers for food design (pp. 279–311). Academic Press. https://doi.org/10.1016/B978-0-12-811449-0.00009-8

Hedrick, V.E., Nieto, C., Grilo, M.F., Sylvetsky, A.C., 2023. Non-sugar sweeteners: Helpful or harmful?: The challenge of developing intake recommendations with the available research. British Medical Journal 9 (383), e075293. Available at https://doi.org/10.1136/bmj-2023-075293.

HHOYA, 2025. Product features: Thaumatin HHOYA. Available at https://hhoya.com/holistic-sustainable-ingredient-solutions/our-products/natural-sweetners/thaumatin, https://doi.org/10.1016/B978-1-63067-044-3.50005-4.

Iko Afé, O.H., Kpoclou, Y.E., Douny, C., Anihouvi, V.B., Igout, A., Mahillon, J., et al., 2021. Chemical hazards in smoked meat and fish. Food Science & Nutrition 9 (12), 6903–6922. Available at https://doi.org/10.1002/fsn3.2633 18.

Imbarex, 2025. Natural dairy colorants: A Growing demand in the food industry imbarex natural colors. Available at https://imbarex.com/natural-dairy-colorants/.

Iommelli, P., Spina, A.A., Vastolo, A., Infascelli, L., Lotito, D., Musco, N., et al., 2025. Functional and economic role of some Mediterranean medicinal plants in dairy ruminants' feeding: A review of the effects of garlic, oregano, and rosemary. Animals (Basel) 15 (5), 657. Available at https://doi.org/10.3390/ani15050657 24.

Jia, F., Ye, K., Zhang, C., Zhang, S., Fu, M., Liu, X., et al., 2022. Effects of vitamin C on the structural and functional characteristics of wheat gluten. Grain & Oil Science and Technology 5 (2), 79–86. Available at https://doi.org/10.1016/j.gaost.2022.04.001.

Kaim, U., Goluch, Z.S., 2023. Health benefits of bread fortification: A systematic review of clinical trials according to the PRISMA statement. Nutrients 15 (20), 4459. Available at https://doi.org/10.3390/nu15204459 20.

Kalsec, 2025. Natural sources of yellow colour extracts Kalsec. Available at https://www.kalsec.com/natural-food-color/insights/natural-yellow-colour-extracts.

Kapadia, P., Newell, A.S., Cunningham, J., Roberts, M.R., Hardy, J.G., 2022. Extraction of high-value chemicals from plants for technical and medical applications. International Journal of Molecular Sciences 23 (18), 10334. Available at https://doi.org/10.3390/ijms231810334.

Kaur, P., Ghoshal, G., Banerjee, U.C., 2019. 3—Traditional bio-preservation in beverages: Fermented beverages. In: Grumezescu, A., Holban, A.M. (Eds.), Preservatives and Preservation Approaches in Beverages. Academic Press, pp. 69–113. Available at https://doi.org/10.1016/B978-0-12-816685-7.00003-3.

Khoo, H.E., Azlan, A., Tang, S.T., Lim, S.M., 2017. Anthocyanidins and anthocyanins: Colored pigments as food, pharmaceutical ingredients, and the potential health benefits. Food & Nutrition Research 61 (1), 1361779. Available at https://doi.org/10.1080/16546628.2017.136177 13.

Khubber, S., Marti-Quijal, F. J., Tomasevic, I., Remize, F., & Barba, F. J. (2022). Lactic acid fermentation as a useful strategy to recover antimicrobial and antioxidant compounds for food and by-products. Current Opinion in Food Science, 43, Article 100795. https://doi.org/10.1016/j.cofs.2021.11.013

Kim, W., Wang, Y., Selomulya, C., 2020. Dairy and plant proteins as natural food emulsifiers. Trends in Food Science & Technology 105, 261–272. Available at https://doi.org/10.1016/j.tifs.2020.09.012.

Kroyer, G.T., 1999. The low calorie sweetener stevioside: Stability and interaction with food ingredients. LWT-Food Science and Technology 32 (8), 509–512. Available at https://doi.org/10.1006/fstl.1999.0585.

Köhlera, K., Schuchmann, H.P., 2011. Homogenisation in the dairy process—Conventional processes and novel techniques. Procedia Food Science 1, 1367–1373. Available at https://doi.org/10.1016/j.profoo.2011.09.202.

Lagouri, V., Boskou, D., 1995. Screening for antioxidant activity of essential oils obtained from spices. In: Charalambous, G. (Ed.), Food Flavors: Generation, Analysis and Process Influence. Elsevier Science B V, pp. 869–879.

Lal, S.N.D., Charmian, J., O'Connor, C.J., Eyres, L., 2006. Application of emulsifiers/stabilizers in dairy products of high rheology. Advances in Colloid and Interface Science 433–437 123–126.

Latoch, A., Czarniecka-Skubina, E., Moczkowska-Wyrwisz, M., 2023. Marinades based on natural ingredients as a way to improve the quality and shelf life of meat: A review. Foods 12 (19), 3638. Available at https://doi.org/10.3390/foods12193638.

List, G.R., 2015. 1 - Soybean lecithin: Food, industrial uses, and other applications. In: Ahmad, M.U., Xu, X. (Eds.), Polar Lipids. Elsevier, pp. 1–33. ISBN9781630670443. Available at https://doi.org/10.1016/B978-1-63067-044-3.50005-4.

Liu, R., Wang, X., Goff, H.D., Cui, S.W., 2024a. Application of yellow mustard gum in preparation of egg-free mayonnaise. International Journal of Food Science and Technology 59 (12), 8972–8982. Available at https://doi.org/10.1111/ijfs.17365.

Liu, Y., Wu, Q., Zhang, J., Yan, W., Mao, X., 2024b. Food emulsions stabilized by proteins and emulsifiers: A review of the mechanistic explorations. International Journal of Biological Macromolecules 261 (Part 1), 129795. Available at https://doi.org/10.1016/j.ijbiomac.2024.129795.

Lourenço, S.C., Martins, M.M., Alves, V.D., 2019. Antioxidants of natural plant origins: From sources to food industry applications. Molecules 24 (22), 4132. Available at https://doi.org/10.3390/molecules24224132 15.

Maleš, I., Pedisić, S., Zorić, Z., Elez-Garofulić, I., Repajić, M., You, L., et al., 2022. The medicinal and aromatic plants as ingredients in functional beverage production. Journal of Functional Foods 96, 105210. Available at https://doi.org/10.1016/j.jff.2022.105210 ISSN1756-4646.

Manessis, G., Kalogianni, A.I., Lazou, T., Moschovas, M., Bossis, I., Gelasakis, A.I., 2020. Plant-derived natural antioxidants in meat and meat products. Antioxidants 9 (12), 1215. Available at https://doi.org/10.3390/antiox9121215.

Mariotti, M., Alamprese, C., 2012. About the use of different sweeteners in baked goods. Influence on the mechanical and rheological properties of the doughs. Lwt—Food Science and Technology 48 (1), 9–15. Available at https://doi.org/10.1016/J.LWT.2012.03.001.

Martinez-Zamore, L., Penalver, R., Ros, G., Nieto, G., 2021. Olive tree derivatives and hydroxytyrosol: Their potential effects on human health and its use as functional ingredient in meat. Foods 10 (11), 2611. Available at https://doi.org/10.3390/foods10112611.

Mashabela, M.N., Ndhlovu, P.T., Mbeng, W.O., 2022. Herbs and spices' antimicrobial properties and possible use in the food sector. Herbs and Spices—New Advances. IntechOpen. Available at https://doi.org/10.5772/intechopen.108143.

McCain, H.R., Kaliappan, S., Drake, M.A., 2018. Invited review: Sugar reduction in dairy products. Journal of Dairy Science 101 (10), 8619–8640. Available at https://doi.org/10.3168/jds.2017-14347.

McClure, B.N., Sebranek, J.G., Kim, Y.H., Sullivan, G.A., 2011. The effects of lactate on nitrosylmyoglobin formation from nitrite and metmyoglobin in a cured meat system. Food chemistry 129 (3), 1072–1079. Available at https://doi.org/10.1016/j.foodchem.2011.05.077.

McKee, L., Latner, T., 2000. Underutilized sources of dietary fiber: A review. Plant Foods for Human Nutrition 55, 285–304. Available at https://doi.org/10.1023/A:1008144310986.

Mitraka, G.C., Kontogiannopoulos, K.N., Batsioula, M., Banias, G.F., Assimpoulou, A.N., 2021. Spent coffee grounds: Valorization towards the recovery of caffeine and chlorogenic acid: A response surface methodology approach. Sustainability 13 (16), 8818. Available at https://doi.org/10.3390/su13168818.

Molina, R.E., Benjamin, M., Bohrer, S.M., Mejia, V., 2023. Phosphate alternatives for meat processing and challenges for the industry: A critical review. Food Research International 166, 112624. Available at https://doi.org/10.1016/j.foodres.2023.112624.

Mordor Intelligence, 2025. Batter and breader premixes market size & share analysis—Growth trends & forecasts (2025—2030). . Available at https://www.mordorintelligence.com/industry-reports/batter-and-breader-premixes-market.

Mudgil, D., Barak, S., Khatkar, B.S., 2014. Guar gum: processing, properties and food applications—A review. Journal of Food Science and Technology 51 (3), 409–418. Available at https://doi.org/10.1007/s13197-011-0522-x.

Mukherjee, S.K., Albury, M.N., Pederson, C.S., Vanveen, A.G., Steinkraus, K.H., 1965. Role of *Leuconostoc Mesenteroides* in leavening the batter of idli, a fermented food of India. Applied Microbiology 13 (2), 227–231. Available at https://doi.org/10.1128/am.13.2.227-231.1965.

Muthu Lakshmi, S., Bharathi, P., Lakshana, R., Soreng, I., Hiaridas, S., 2025. Revolutionizing of food preservation with vitamin C: Applications and impacts of vitamin C in food products. International Journal of Science, Engineering and Management (IJSEM 12 (04), 1–12 ISSN(Online).

Nabi, B.G., Mukhtar, K., Ahmed, W., Manzoor, M.F., Ranjha, M.A.N., Bhat, K.M., et al., 2023. Natural pigments: Anthocyanins, carotenoids, chlorophylls, and betalains as colorants in food products. Food Bioscience 52, 102403. Available at https://doi.org/10.1016/j.fbio.2023.102403.

Naik, V., Poyil, T., 2022. Application of stevia (*Stevia rebaudiana* Bertoni.) in food products. The Pharma Innovation Journal 11 (7), 2056–2060.

Natural, A., 2025. Natural Extracts: Transforming the Beverage Industry Arjuna Natural. Available at https://arjunanatural.com/natural-extracts-transforming-the-beverage-industry/#:~:text=Flavor%20Enhancement:%20Natural%20extracts%20are,but%20also%20offer%20health%20benefits.

Neves, M.I.N., Silva, E.K., Meireles, M.A.A., 2021. Natural blue food colorants: Consumer acceptance, current alternatives, trends, challenges, and future strategies. Trends in Food Science & Technology 112, 163–173. Available at https://doi.org/10.1016/j.tifs.2021.03.023.

NIH, 2017. National Cancer Institute. Meat cooked at high temperatures and cancer risk. Available at https://www.cancer.gov/about-cancer/causes-prevention/risk/diet/cooked-meats-fact-sheet#what-research-is-being-conducted-on-the-relationship-between-the-consumption-of-hcas-and-pahs-and-cancer-risk-in-humans.

Novais, C., Adriana, K., Molina, R.M., Abreu, V., Santo-Buelga, C., Ferreira, I.C.F.R., et al., 2022. Natural food colorants and preservatives: A review, a demand, and a challenge. Journal of Agricultural and Food Chemistry 70 (9), 2789–2805. Available at https://doi.org/10.1021/acs.jafc.1c07533.

Ouyang, Z., You, K., Mi, M., Lin, Y., Zheng, S., 2025. Microbial production of purine nucleotides, nucleosides, and nucleobases: Advances and perspectives. Biotechnology journal 20 (9), e70115. Available at https://doi.org/10.1002/biot.70115.

Pandya, Y., Bakshi, M., Sharma, A., 2022. Agar-agar extraction, structural properties and applications: A review. The Pharma Innovation Journal; SP 11 (6), 1151–1157.

Panseri, S., Arioli, F., Cristina, B., Giacomo, M., Pavlovic, R., Chiesa, L.M., 2020. Detection of polyphosphates in seafood and its relevance toward food safety. Food chemistry 332, 127397. Available at https://doi.org/10.1016/j.foodchem.2020.127397.

Plant lipids, 2025. Natural food colours. . Available at https://plantlipids.com/natural-food-colours/#:~:text=Explore%20our%20vibrant%20pink%20and,to%20meet%20your%20specific%20needs.

Pool, H., Quintanar, D., Figueroa, J.D., Mano, C.M., Etelvino, J.H., Bechara, L.A., et al., 2012. Antioxidant effects of quercetin and catechin encapsulated into PLGA nanoparticles. Journal of Nanomaterials 2012 (1), 145380.

Poyil, T., Naik, V., Sajith, K.S., 2022. Mustard based products the pharma innovation. The Journal 11 (7), 2061–2065.

Procool, 2024. Drink preservatives: Essential solutions for the beverage industry Procool. Available at https://procoolmfg.com/drink-preservatives/#toc.

Putta, S., 2020. FSSAI guidance and notification on nutraceuticals—An insight. Available at https://fssai.gov.in/upload/media/FSSAI_News_Guidance_FNB_09_06_2020.pdf accessed 06.08.2025.

Rao, K.J., Pagote, C.N., 2018. Use of artifcial sweeteners in indian traditional dairy products. Ṫ. Food Nutr J 177 (3). Available at https://doi.org/10.29011/2575-7091.100077.

Rajni (2023) Gum Arabic: Applications in Industries and Benefits for Human Health. Just Agriculture 4(3). 136-141. (e-ISSN: 2582-8223)

Reese, A.T.Madden, A.A.Joossens, M.Lacaze, G.Dunn, R.R. Influences of ingredients and bakers on the bacteria and fungi in sourdough starters and bread. Available at mSphere5:10.1128/msphere.00950-19.

Reichembach, L.H., Petkowicz, C.L., 2021. Pectins from alternative sources and uses beyond sweets and jellies: An overview. Food Hydrocolloids 118, 106824. Available at https://doi.org/10.1016/j.foodhyd.2021.106824.

Reshna, K.R., Sreerag, G., Balakrishnan, P., 2022. Introduction to flavor and fragrance in food processing. Flavors and Fragrances in Food Processing: Preparation and Characterization Methods. *ACS Symposium Series*, pp. 1–19. 1433. Available at https://doi.org/10.1021/bk-2022-1433.ch001.

Rodríguez, M., Caleja, C., Núñez, E.B., Pereira, E., Corral, M., Reis, F., et al., 2021. Flavonoids: A group of potential food additives with beneficial health effects. In: Prieto, M.A., Otero, P. (Eds.), IntechOpen. Available at https://doi.org/10.5772/intechopen.101466.

Roy, V.C., Arachchige, W., Chamika, S., Park, J.-S., Ho, T.C., Khan, F., et al., 2021. Preparation of bio-functional surimi gel incorporation of fish oil and green tea extracts:

Physico-chemical activities, in-vitro digestibility, and bacteriostatic properties. Food control 130, 108402. Available at https://doi.org/10.1016/j.foodcont.2021.108402.
Roy, S., Deshmukh, R.K., Tripathi, S., Gaikwad, K.K., Das, S.S., Sharma, D., 2023. Recent advances in the carotenoids added to food packaging films: A review. Foods 12 (21), 4011. Available at https://doi.org/10.3390/foods12214011.
Saraiva, A., Carrascosa, C., Raheem, D., Ramos, F., Raposo, A., 2020. Natural sweeteners: The relevance of food naturalness for consumers, food security aspects, sustainability and health impacts. International journal of environmental research and public health 17 (17), 6285. Available at https://doi.org/10.3390/ijerph17176285.
Scappaticci, G., Mercanti, N., Pieracci, Y., Ferrari, C., Mangia, R., Marianelli, A., et al., 2024. Bread improvement with nutraceutical ingredients obtained from food by-products: Effect on quality and technological aspects. Foods 13 (6), 825. Available at https://doi.org/10.3390/foods13060825.
Selahvarzi, A., Ramezan, Y., Sanjabi, M.R., et al., 2021. Investigation of antimicrobial activity of orange and pomegranate peels extracts and their use as a natural preservative in a functional beverage. J.Food Measurement and Characterization. Available at https://doi.org/10.1007/s11694-021-01141-z.
Sengar, G., Sharma, H.K., 2014. Food caramels: A review. Journal of Food Science and Technology 51 (9), 1686–1696. Available at https://doi.org/10.1007/s13197-012-0633-z.
Sezgin, A., Ayyıldız, S., 2017. Food Additives: Colorants. Available at https://www.researchgate.net/publication/322404775_Food_Additives_Colorants.
Shakil, M.H., Trisha, A.T., Rahman, M., Talukdar, S., Kobun, R., Huda, N., et al., 2022. Nitrites in cured meats, health risk issues, alternatives to nitrites: A review. Foods 11 (21), 3355. Available at https://doi.org/10.3390/foods11213355 25.
Sharma, S., Wani, K.M., Mujahid, S.M., Jayan, L.S., Rajan, S.R., 2026. Review on pectin: Sources, properties, health benefits and its applications in food industry. Journal of Future Foods 6 (2), 205–219. Available at https://doi.org/10.1016/j.jfutfo.2024.04.009.
Silva Leal, M.R., Albuquerque, P.B.S., Rodrigues, N.E.R., Silva, P.M.S., de Oliveira, W.F., Correia, M.T.S., et al., 2024. A review on the use of polysaccharides as thickeners in yogurts. Carbohydrate Polymer Technologies and Applications 8, 100547. Available at https://doi.org/10.1016/j.carpta.2024.100547.
Silva, M.M., Reboredo, F.H., Lidon, F.C., 2022. Food colour additives: A synoptical overview on their chemical properties, applications in food products, and health side effects. Foods 11 (3), 379. Available at https://doi.org/10.3390/foods11030379.
Singh, A.P., Kumar, S., 2019. Applications of tannins in industry. Tannins—Structural properties, biological properties and current knowledge. IntechOpen. Available at https://doi.org/10.5772/intechopen.85984.
Singh, T., Pandey, V.K., Dash, K.K., Zanwar, S., Singh, R., 2023. Natural bio-colorant and pigments: Sources and applications in food processing. Journal of Agriculture and Food Research 12, 100628. Available at https://doi.org/10.1016/j.jafr.2023.100628.
Singh, N., Sudha, M.L., 2024. Natural food flavours: A healthier alternative for bakery industry—a review. Journal of Food Science and Technology 61 (2024), 642–650. Available at https://doi.org/10.1007/s13197-023-05782-4.
Siwach, R., Tokas, J., Seth, R., 2016. Use of lycopene as a natural antioxidant in extending the shelf-life of anhydrous cow milk fat. Food Chemistry 199, 541–546. Available at https://doi.org/10.1016/j.foodchem.2015 12.009.
Soni, R.K., Soni, R., Meena, M.L., Kumar, R., Kumar, J., Azad, A.K., et al., 2024. Comparative study on the effect of citric acid and lactic acid coagulation on paneer quality. International Journal of Advanced Biochemistry Research 2024; SP 8 (7), 615–617.

de Souza, E.L., da Cruz, A.E.T., de Sousa, G.J.P., 2019. 5 - Emerging nonchemical potential antimicrobials for beverage preservation. In: Grumezescu, A.M., Holban, A.M. (Eds.), Preservatives and Preservation Approaches in Beverages. Academic Press, pp. 149–178. ISBN9780128166857. Available at https://doi.org/10.1016/B978-0-12-816685-7.00005-7.

Stampfi, L., Nersten, B., 1995. Emulsifiers in bread making. Food Chemistry 52, 353–360.

Sun, H., Wang, Y., He, Y., Liu, B., Mou, H., Chen, F., et al., 2023. Microalgae-derived pigments for the food industry. Marine drugs 21 (2), 82. Available at https://doi.org/10.3390/md21020082 2023 Jan 25.

Swain, M., Ray, R., Patra, J.K., 2012. Citric acid: Microbial production and applications in food and pharmaceutical industries. In: Vargas, D.A., Medina, J.V. (Eds.), Citric Acid: Synthesis, Properties and Applications. Nova Science Publisher, pp. 97–118.

Sylvetsky, A.C., Rother, K.I., 2016. Trends in the consumption of low-calorie sweeteners. Physiology & Behavior 164 (Pt B), 446–450. Available at https://doi.org/10.1016/j.physbeh.2016.03.030.

Tabanty, Z.G., Tenyang, N., Birault, L., Kermarrec, A., Gacel, A., Kansci, G., et al., 2023. Effect of some local plant extracts on fatty acid composition of fish (*Alestes baremoze*) during smoking and sun drying in the Far-North region of Cameroon. Food Science & Nutrition 11 (9), 5621–5637. Available at https://doi.org/10.1002/fsn3.3636 16.

Tian, X., Zhong, F., Xia, Y., 2022. Dynamic characteristics of sweetness and bitterness and their correlation with chemical structures for six steviol glycosides. Food Research International 151, 110848. Available at https://doi.org/10.1016/j.foodres 2021.110848.

Udo, T., Gopinath, M., Mohan, A., Singh, R.K., Kong, F., 2023. Current and emerging applications of carrageenan in the food industry. Food Research International 173 (Part 2), 113369. Available at https://doi.org/10.1016/j.foodres.2023.113369.

Ullrich, T., 2025. The ins and outs of building an effective nutraceutical supply chain strategy. Tips For Building A Strong Nutraceutical Supply Chain Strategy.

Vandamme, E.J., 1992. Production of vitamins, coenzymes and related biochemicals by biotechnological processes. Journal of Chemical Technology and Biotechnology 53 (4), 313–327. Available at https://doi.org/10.1002/jctb.280530402 PMID:1368195.

Veenstra, J.P., Johnson, J.J., 2019. Oregano (*Origanum vulgare*) extract for food preservation and improvement in gastrointestinal health. Int J Nutr 3 (4), 43–52. Available at https://doi.org/10.14302/issn.2379-7835.ijn-19-2703.

Vega, E.N., Ciudad-Mulero, M., Fernández-Ruiz, V., Barros, L., Morales, P., 2023. Natural sources of food colorants as potential substitutes for artificial additives. Foods 12 (22), 4102. Available at https://doi.org/10.3390/foods12224102 12.

Venkatesan, U., Muniyan, R., 2024. Review on the extension of shelf life for fruits and vegetables using natural preservatives. Food science and biotechnology 33 (11), 2477–2496. Available at https://doi.org/10.1007/s10068-024-01602-3 18.

Vesta~VAR, 2025. Chilled fish technologies. Food additives (preservatives) Varex-6 and Varex-7. Vesta~VAR. Available at https://www.vesta-var.ru/en/products/for-chilled-fish/#:~:text=Safe%20food%20additive%20Varex%2D7,Further%20reading.

Vignesh, A., Amal, T.C., Sarvalingam, A., Vasanth, K., 2024. A review on the influence of nutraceuticals and functional foods on health. Food Chemistry Advances 5, 100749. Available at https://doi.org/10.1016/j.focha.2024.100749.

Virginia Cooperative Extension, 2020. Reduced oxygen packaging and food safety concerns in seafood Bulletin#FST-358NP. Virginia Tech. Virginia State University 5pp.

Walayat, N., Xiong, H., Xiong, Z., Moreno, H.M., Nawaz, A., Niaz, N., et al., 2022. Role of cryoprotectants in surimi and factors affecting surimi gel properties: A review. Food Reviews

International 38 (6), 1103–1122. Available at https://doi.org/10.1080/87559129.2020.1768403.

Wang, Z., Wang, L., Yu, X., Wang, X., Zheng, Y., Hu, X., et al., 2024. Effect of polysaccharide addition on food physical properties: A review. Food chemistry 431, 137099. Available at https://doi.org/10.1016/j.foodchem.2023.137099.

WHO (World health organization). (2023). Food additives https://www.who.int/news-room/fact-sheets/detail/food-additives accessed 06.08.2025.

Wu, Y., Eskin, N.A.M., Cui, W., Pokharel, B., 2015. Emulsifying properties of water soluble yellow mustard mucilage: A comparative study with gum Arabic and citrus pectin. Food Hydrocolloids Volume 47, 191–196. Available at https://doi.org/10.1016/j.foodhyd.2015.01.020.

Wu, T., Zhu, W., Chen, L., Jiang, T., Dong, Y., Wang, L., et al., 2023. A review of natural plant extracts in beverages: Extraction process, nutritional function, and safety evaluation. Food Research International 172. Available at https://doi.org/10.1016/j.foodres.2023.113185.

Ran, X.-L., Zhang, M., Wang, Y., Adhikari, B., 2019. Novel technologies applied for recovery and value addition of high value compounds from plant byproducts: A review. Critical Reviews in Food Science and Nutrition 59 (3), 450–461. Available at https://doi.org/10.1080/10408398.2017.1377149.

Xie, A., Dong, Y., Liu, Z., Li, Z., Shao, J., Li, M., et al., 2023. A review of plant-based drinks addressing nutrients, flavor, and processing technologies. Foods 12 (21), 3952. Available at https://doi.org/10.3390/foods12213952.

Yan, M.R., Welch, R., Rush, E.C., Xiang, X., Wang, X., 2019. A sustainable wholesome foodstuff; health effects and potential dietotherapy applications of yacon. Nutrients 11 (11), 2632. Available at https://doi.org/10.3390/nu11112632 3.

Yvonne, M., Jideani, V., 2018. Factors affecting the stability of emulsions stabilised by biopolymers. In: Science and Technology Behind Nanoemulsions. InTech Open. Available at http://dx.doi,org/10.5772/intechopen.75308.

Zakynthinos, G., Varzakas, T., 2016. Carotenoids: From plants to food industry. Curr Res Nutr. Food Sci 2016 4 (Special Issue Carotenoids March 2016). Available at https://doi.org/10.12944/CRNFSJ.4.Special-Issue1.04.

Zanjani, M.M., Yousefi, M., Ehsani, A., 2019. Challenges and approaches for production of a healthy and functional mayonnaise sauce. Food Science & Nutrition 7 (8), 2471–2484. Available at https://doi.org/10.1002/fsn3.1132 18.

Zeece, M., 2020. Chapter 6 – Flavors. In: Zeece, M. (Ed.), Introduction to the chemistry of food. Academic Press, pp. 213–250. ISBN9780128094341. Available at https://doi.org/10.1016/B978-0-12-809434-1.00006-2.

Zhu, F., 2024. Modifications of whey proteins for emulsion based applications: Current status, issues and prospectives. Food Research International 178, 113935. Available at https://doi.org/10.1016/j.foodres.2024.113935.

Safety and quality control of natural and nutraceutical food additives

13

Velumurugan Shanmugam[1,2] *and Umalatha*[3]
[1]Business Development & Quality Affairs, Biogenero Labs Pvt. Ltd., Bengaluru, Karnataka, India, [2]Phytosol India Pvt. Ltd., Bengaluru, Karnataka, India, [3]Department of Food Science and Technology, Padmashree Institute of Management and Sciences, Bengaluru, Karnataka, India

13.1 Introduction

Food additives are integral to modern food systems, contributing to product safety, stability, quality, and consumer acceptability. In recent years, growing consumer demand for clean-label and health-oriented foods has accelerated the use of natural and nutraceutical food additives as alternatives to conventional synthetic additives. These additives are commonly derived from plant, animal, microbial, marine, or mineral sources and may offer both technological functionality and additional nutritional or physiological benefits.

Despite their natural origin, such additives cannot be assumed to be inherently safe. Standard food safety and food chemistry literature emphasizes that natural and nutraceutical additives often exhibit compositional complexity and variability, influenced by factors such as raw material source, environmental conditions, and processing methods (Branen et al., 2002; Damodaran et al., 2007). This variability necessitates standardization and specification limits to ensure consistent safety and functional performance across batches.

International food safety frameworks have consistently highlighted that all food additives, irrespective of origin, require systematic safety evaluation based on scientific evidence. International standards such as the Codex General Standard for Food Additives (GSFA) provide harmonized guidelines on permitted additives, usage levels, and safety evaluation criteria (Codex Alimentarius Commission, 2023). Guidance developed under the auspices of the Codex Alimentarius Commission and the World Health Organization underscores that "natural" status does not exempt an additive from toxicological assessment, exposure evaluation, and quality control requirements. Consequently, ensuring the safety and quality of natural and nutraceutical food additives demands an integrated, science-based approach covering the entire value chain.

This chapter, therefore, focuses on the critical safety risks, quality control measures, and regulatory considerations associated with natural and nutraceutical

Nutraceutical Applications in Food Processing and Preservation. DOI: https://doi.org/10.1016/B978-0-443-45784-5.00002-2

food additives, drawing upon established food science principles and internationally accepted regulatory guidance.

13.2 Sources of natural and nutraceutical food additives

Natural and nutraceutical food additives originate from diverse biological and mineral sources. Many of these additives are already permitted for food use under national and international regulations, while a substantial number of underutilized or unexplored sources continue to attract scientific and industrial interest. However, before any new source can be introduced into the food supply, a comprehensive evaluation of identity, safety, toxicity, stability, and functional performance is essential, as emphasized in food additive risk-assessment frameworks developed by the Joint FAO/WHO Expert Committee on Food Additives.

Each source category presents distinct safety and quality challenges, making source-specific evaluation essential for the safe application of natural and nutraceutical food additives.

13.2.1 *Plant-derived additives*

Plant-derived additives constitute the largest group of natural food additives and include colorants, antioxidants, flavoring agents, sweeteners, and stabilizers. Widely used examples such as curcumin, carotenoids, polyphenols, and plant gums are valued for both their technological functions and nutraceutical properties. Many of these additives are well established and included in international standards or national regulations.

Nevertheless, plant-derived additives may pose safety concerns related to botanical misidentification, the presence of inherent toxic constituents, pesticide residues, and variability in bioactive composition. Authoritative reviews and regulatory opinions emphasize the importance of correct botanical authentication, standardized extraction methods, and toxicological evaluation to ensure safety and reproducibility (EFSA, 2012; Codex Alimentarius Commission, 2023).

13.2.2 *Animal-derived additives*

Animal-derived food additives, such as gelatin, caseinates, and certain enzymes, are primarily used for their functional properties, including gelling, emulsification, and texture modification. These additives are generally well characterized and regulated when sourced and processed appropriately.

However, potential safety risks include microbiological contamination, residues of veterinary drugs, and allergenic potential. Food safety authorities require stringent control of raw material sourcing, hygienic processing, and compliance with residue limits to minimize risks to consumers (FDA, 2022; EFSA, 2011). Ethical considerations and dietary preferences may also influence acceptance, underscoring the need for transparency and compliance.

13.2.3 Microbial and fermentation-derived additives

Microbial and fermentation-derived additives include organic acids, enzymes, polysaccharides, and other metabolites produced through controlled fermentation processes. These additives are increasingly favored due to better process control, scalability, and consistency compared with many plant-derived ingredients.

Despite these advantages, safety concerns may arise from inappropriate strain selection, formation of undesirable by-products, or insufficient downstream purification. International guidance clearly states that only well-characterized, nonpathogenic microbial strains should be used for food applications, and that fermentation processes must be adequately validated to ensure product safety (WHO and FAO, 1987; ECFA, 2006). Microbial strains used in food applications are often required to meet safety criteria such as Qualified Presumption of Safety (QPS) in the European Union or Generally Recognized as Safe (GRAS) status in the United States.

13.2.4 Marine-derived additives

Marine-derived additives, including polysaccharides, pigments, and bioactive lipids, offer unique functional and nutraceutical properties. However, marine environments are susceptible to contamination by heavy metals and naturally occurring marine toxins such as domoic acid, saxitoxin, and okadaic acid, which may bioaccumulate in raw materials.

Food safety literature highlights the importance of careful source selection, monitoring of environmental contaminants, and effective purification processes to ensure the safety of marine-derived additives before their incorporation into food products (EFSA, 2012; Codex Alimentarius Commission, 2023).

13.2.5 Mineral-based additives

Mineral-based additives are commonly used for fortification and functional purposes. Their safety depends on chemical form, bioavailability, dosage, and cumulative dietary exposure. Excessive intake or inappropriate formulation may result in adverse health effects, particularly in vulnerable population groups.

Regulatory authorities specify Acceptable Daily Intake (ADI) levels and usage conditions for mineral additives, reinforcing the need for controlled application and strict compliance with established standards (JECFA, 2019).

13.3 Safety risks in the value chain

Natural and nutraceutical food additives cannot be used directly in food systems without appropriate processing and evaluation. Safety risks associated with these additives arise at multiple stages of the value chain, beginning from raw material sourcing and extending through extraction, processing, formulation, and packaging. Unlike synthetic additives produced under tightly controlled industrial conditions, natural additives are often influenced by environmental, agricultural, and handling

Figure 13.1 Safety risks in the value chain of natural and nutraceutical food additives. Potential safety risks occurring across the value chain of natural and nutraceutical food additives, from raw material collection and extraction to formulation, packaging, and storage.

factors, which may significantly affect their safety profile (Branen et al., 2002; Damodaran et al., 2007) (Fig. 13.1).

13.3.1 Raw material collection and storage

Safety risks may originate at the stage of raw material collection, particularly when plant, animal, or marine sources are harvested under variable environmental conditions. Factors such as soil quality, water contamination, agricultural inputs, and postharvest handling practices can influence the presence of contaminants and degradation products in raw materials. Improper drying, storage, or transportation can further promote microbial growth or chemical deterioration, thereby compromising safety (Rahman, 2007).

Food safety literature consistently indicates that good agricultural and collection practices are essential for minimizing contamination risks at the source. Traceability and documentation of origin are therefore critical components of safety assurance for natural and nutraceutical additives (Branen et al., 2002).

13.3.2 Extraction and processing

Extraction and processing steps represent a critical control point for safety. The choice of extraction method, solvents, temperature, and processing time can influence both the concentration of bioactive compounds and the presence of undesirable residues. Inadequate process control may lead to the concentration of toxic constituents or degradation of beneficial components (Shahidi, 2012).

Furthermore, improper purification can result in residual solvents or process-related contaminants. Scientific reviews consistently highlight the importance of validated extraction protocols and downstream processing to ensure reproducible quality and safety of natural food additives (Shahidi and Ambigaipalan, 2015; Damodaran et al., 2007).

13.3.3 Formulation and packaging

Safety risks may also emerge during formulation and packaging. Interactions between nutraceutical additives and food matrices can alter stability, bioavailability, or toxicity. Packaging materials may influence oxidative stability or lead to the migration of undesirable compounds if not properly selected.

Food safety and processing literature stresses that formulation design and packaging selection must be evaluated together to preserve both functional performance and safety throughout the product's shelf life (Rahman, 2007).

Unlike synthetic additives, where risks are largely confined to manufacturing control, the safety of natural and nutraceutical additives is inherently distributed across the entire value chain, making upstream interventions as critical as downstream regulatory oversight.

13.4 Contamination and residues

Contamination remains one of the most significant safety challenges associated with natural and nutraceutical food additives. Due to their biological origin, these additives may be exposed to multiple sources of contamination during cultivation, harvesting, and processing.

13.4.1 Heavy metals and minerals

Heavy metals such as lead, cadmium, arsenic, and mercury may be introduced into natural additives through contaminated soil, water, or marine environments. These elements can bioaccumulate, particularly in plant and marine-derived additives, posing potential health risks when consumed over time. Food safety texts emphasize the necessity of monitoring and controlling heavy metal levels to ensure consumer safety (Branen et al., 2002; Rahman, 2007). Regulatory limits for contaminants and additives are defined by international standards such as Codex GSFA and national authorities.

13.4.2 Pesticides and veterinary drug residues

Residues of pesticides and veterinary drugs may be present in raw materials derived from agricultural or animal sources. Inadequate adherence to good agricultural and veterinary practices can result in residue carryover into extracted additives. Scientific reviews highlight that residue monitoring and compliance with established limits are essential components of quality control for natural additives (Damodaran et al., 2007).

13.4.3 Mycotoxins and marine toxins

Mycotoxins produced by fungi and toxins of marine origin represent serious safety hazards. Plant-derived additives may be contaminated by mycotoxins if raw materials are improperly stored, while marine additives may accumulate naturally occurring toxins depending on environmental conditions. These contaminants are of particular concern due to their stability and potent biological effects (Shahidi, 2012).

13.4.4 Residual solvents

Residual solvents from extraction processes can pose toxicological concerns if not adequately removed. The presence of such residues highlights the importance of selecting appropriate solvents and implementing effective purification and validation steps, as emphasized in food processing and safety literature (Rahman, 2007). Residual solvents are classified based on toxicity (e.g., ICH Class 1, Class 2, and Class 3 solvents), and their levels must comply with established regulatory limits. Although originally developed for pharmaceutical applications, this classification provides a useful framework for assessing solvent safety in food-related processes.

13.5 Processing and interaction risks

Processing conditions play a decisive role in determining the safety and functional integrity of natural and nutraceutical food additives. Unlike many synthetic additives that exhibit high thermal and chemical stability, bioactive compounds of natural origin are often sensitive to processing stresses such as heat, light, oxygen, and mechanical shear. These factors may lead to degradation, loss of functional efficacy, or formation of undesirable reaction products (Shahidi and Ambigaipalan, 2015; Damodaran et al., 2007).

13.5.1 Effects of heat, light, and oxygen

Thermal processing is widely employed in food manufacturing to ensure safety and shelf stability; however, excessive heat exposure can degrade heat-labile nutraceutical compounds such as polyphenols, carotenoids, and certain vitamins. Oxidative reactions induced by heat and oxygen may result in the formation of degradation products with altered safety and sensory properties. Processing may also lead to the formation of reaction products such as Maillard reaction intermediates and lipid oxidation products, which may alter both safety and functional properties. Light exposure, particularly ultraviolet and visible light, can further accelerate photodegradation of pigments and antioxidants (Damodaran et al., 2007). Processing may also lead to formation of reaction products such as Maillard reaction intermediates or oxidation products, which may alter safety and functionality.

Food processing literature emphasizes the importance of optimizing processing parameters to balance microbial safety with preservation of bioactive integrity. Protective strategies such as controlled atmosphere processing, antioxidant incorporation, and appropriate packaging selection are therefore essential to minimize processing-induced risks (Rahman, 2007).

13.5.2 Additive–food matrix interactions

The safety and functionality of nutraceutical additives are strongly influenced by their interactions with the food matrix. Components such as proteins, lipids, carbohydrates, and minerals can affect the stability, bioavailability, and reactivity of bioactive compounds. In some cases, matrix interactions may reduce bioactivity through binding or complex formation, while in others they may enhance stability or modify release characteristics, thereby influencing their physiological efficacy (Shahidi, 2012).

From a safety perspective, unintended interactions may alter the chemical nature of additives or promote the formation of secondary reaction products during storage and processing. Understanding matrix-dependent behavior is therefore critical for predicting additive performance and ensuring consistent safety outcomes (Damodaran et al., 2007).

13.5.3 Additive–additive interactions

Natural and nutraceutical food additives are rarely used in isolation. In multi-component food formulations, interactions between additives may influence stability, efficacy, and safety. Synergistic antioxidant effects may be beneficial, whereas antagonistic or reactive interactions could compromise functionality or lead to degradation.

Food formulation studies highlight the need for compatibility assessment when combining multiple natural additives, particularly in complex food systems where pH, moisture, and processing conditions vary widely (Branen et al., 2002).

This sensitivity of bioactive compounds to processing conditions underscores the need to design processing strategies around additive stability rather than treating nutraceuticals as interchangeable with conventional additives.

13.6 Toxicity and allergens

Toxicological evaluation is a cornerstone of safety assessment for natural and nutraceutical food additives. Although these additives are derived from natural sources, they may exhibit dose-dependent toxicity, allergenic potential, or long-term effects that are not fully understood at the time of initial application.

13.6.1 Dose-dependent toxicity

The principle that “the dose makes the poison” applies equally to natural and synthetic substances. Several naturally occurring bioactive compounds may exert adverse effects at high intake levels or under chronic exposure conditions. Food safety texts and toxicological reviews emphasize that safety assessment must consider realistic consumption patterns, cumulative exposure, and vulnerable population groups (Branen et al., 2002; WHO and FAO, 1987). Toxicological evaluation includes determination of NOAEL (No Observed Adverse Effect Level), which forms the basis for establishing ADI.

Inadequate understanding of dose–response relationships may lead to unintended health risks, particularly when nutraceutical additives are incorporated into frequently consumed foods.

13.6.2 Understudied bioactive compounds

A significant challenge in the use of nutraceutical food additives is the limited toxicological data available for many underutilized or novel bioactive compounds. While traditional use may suggest safety, scientific validation through controlled toxicological studies is essential to confirm the absence of adverse effects.

Scientific reviews stress that insufficient characterization of bioactive composition, metabolites, and long-term exposure outcomes represents a major knowledge gap in nutraceutical safety assessment (Shahidi, 2012; Shahidi and Ambigaipalan, 2015).

13.6.3 Allergens from plant, animal, or microbial sources

Allergenic potential represents an important safety consideration for natural food additives. Proteins derived from plant, animal, or microbial sources may trigger allergic reactions in sensitive individuals. Even trace levels of allergenic components can pose risks to vulnerable consumers.

Food allergen management guidelines highlight the need for thorough assessment of allergenicity, clear labeling, and appropriate risk communication to protect consumers while maintaining product integrity (Branen et al., 2002; EFSA, 2011).

13.6.4 Vulnerable population groups

Certain population groups, including infants, pregnant women, elderly individuals, and those with compromised health, may be more susceptible to adverse effects from bioactive food additives. Safety evaluations must therefore consider differential sensitivity and exposure scenarios to ensure broad consumer protection.

Risk-based approaches outlined in food safety literature emphasize the importance of accounting for vulnerable populations during safety assessment and regulatory decision-making (WHO and FAO, 1987).

Given these potential toxicological and allergenic risks, robust quality control systems are essential to ensure that natural and nutraceutical food additives consistently meet safety and performance requirements.

13.7 Quality control of natural and nutraceutical food additives

Quality control is a critical component in ensuring the safety, functionality, and reproducibility of natural and nutraceutical food additives. Unlike synthetic additives, which are typically produced under standardized industrial conditions, natural additives often exhibit variability due to differences in raw material source, processing conditions, and storage environments. Effective quality control systems are therefore essential to ensure consistent performance and consumer safety (Branen et al., 2002; Damodaran et al., 2007).

Quality control of natural and nutraceutical additives encompasses verification of identity and authenticity, standardization of active compounds, assurance of batch-to-batch consistency, and evaluation of functional performance in food systems.

13.7.1 Identity and authenticity

Accurate identification and authentication of natural food additives are fundamental to quality assurance. Misidentification of botanical or biological sources can result in the use of incorrect or potentially toxic materials. Adulteration or substitution with cheaper synthetic compounds further complicates quality assurance.

Analytical techniques such as chromatographic profiling, spectroscopic methods, and fingerprinting approaches are commonly recommended in food quality literature to confirm the identity and authenticity of natural additives (Shahidi, 2012). Proper documentation and traceability systems further support authenticity verification across the supply chain.

13.7.2 Standardization of active compounds

Standardization refers to ensuring consistent levels of key bioactive constituents responsible for functional and nutraceutical effects. Variability in active compound concentration can lead to inconsistent technological performance and unpredictable safety outcomes.

Food science reviews emphasize that standardization requires well-defined specifications, validated analytical methods, and controlled processing conditions. Without such measures, natural additives may fail to meet functional requirements or exceed safe exposure levels (Shahidi and Ambigaipalan, 2015).

13.7.3 Batch-to-batch consistency

Batch-to-batch consistency is particularly challenging for natural and nutraceutical additives due to inherent variability in raw materials. Differences in harvest season, geographic origin, and processing parameters may affect composition and functionality.

Quality management systems described in food additive literature recommend routine monitoring of critical quality attributes and implementation of corrective actions to minimize variability and ensure reliable performance (Branen et al., 2002).

13.7.4 Functional performance evaluation

Beyond chemical composition, natural and nutraceutical additives must be evaluated for their functional performance within specific food matrices. Properties such as antioxidant efficacy, emulsification capacity, color stability, and antimicrobial activity must be validated under realistic processing and storage conditions (Table 13.1).

Functional testing ensures that quality control extends beyond compositional compliance to practical effectiveness in food applications (Damodaran et al., 2007).

Table 13.1 Key quality control parameters for natural and nutraceutical food additives.

Quality aspect	Key considerations	Purpose
Identity	Botanical/biological authentication	Prevent misidentification and substitution
Purity	Absence of contaminants and adulterants	Ensure consumer safety
Standardization	Consistent bioactive content	Reliable functionality and safety
Consistency	Batch-to-batch uniformity	Product reproducibility
Performance	Functional efficacy in food systems	Technological reliability

For natural and nutraceutical food additives, quality control thus functions not merely as a compliance tool, but as a primary risk-mitigation mechanism bridging scientific variability and regulatory expectations.

13.8 Stability and shelf-life

Stability and shelf-life are critical determinants of both safety and functionality of natural and nutraceutical food additives. Many bioactive compounds are inherently sensitive to environmental factors such as temperature, light, oxygen, and moisture, which can lead to degradation during processing and storage (Rahman, 2007).

13.8.1 Bioactive stability

Chemical degradation of bioactive compounds may result in loss of functional efficacy or formation of undesirable degradation products. Oxidation, hydrolysis, and photodegradation are common mechanisms affecting the stability of natural additives.

Food preservation literature emphasizes the importance of stability testing under controlled and accelerated conditions to predict shelf-life and identify critical degradation pathways (Damodaran et al., 2007).

13.8.2 Functional and sensory retention

Stability assessment must consider not only chemical integrity but also retention of functional and sensory attributes. Loss of color intensity, flavor potency, or antioxidant activity can compromise product quality and consumer acceptance.

Ensuring functional and sensory retention throughout shelf life is therefore a key objective of quality control strategies for nutraceutical additives (Shahidi, 2012).

13.8.3 Influence of packaging

Packaging plays a significant role in preserving the stability of natural food additives. Barrier properties against oxygen, moisture, and light can substantially influence degradation rates. Selection of appropriate packaging materials is therefore integral to shelf-life management (Rahman, 2007).

13.9 Adulteration and fraud

Adulteration and fraud represent significant challenges in the supply chain of natural and nutraceutical food additives. Economic incentives may lead to substitution with synthetic analogs, dilution with inactive materials, or mislabeling of origin and composition.

Such practices compromise safety, efficacy, and consumer trust. Food authenticity literature highlights the importance of robust authentication methods and supply chain transparency to detect and prevent fraudulent practices (Shahidi, 2012).

13.9.1 Substitution and dilution

Substitution of natural additives with cheaper synthetic compounds or dilution with inert materials can result in misleading labeling and unpredictable safety outcomes. These practices may also interfere with functional performance in food systems.

13.9.2 Mislabeling and traceability issues

Mislabeling of botanical source, geographic origin, or processing method undermines regulatory compliance and consumer confidence. Effective traceability systems are essential to address these risks and support quality assurance.

13.9.3 Authentication methods

Analytical fingerprinting, compositional profiling, and traceability documentation are commonly recommended approaches to authenticate natural food additives and detect adulteration (Branen et al., 2002; Shahidi, 2012) (Table 13.2).

Table 13.2 Common forms of adulteration and quality risks in natural food additives.

Type of risk	Description	Potential impact
Substitution	Replacement with synthetic or cheaper additives	Safety and regulatory noncompliance
Dilution	Mixing with inert or low-quality materials	Reduced functionality
Mislabeling	Incorrect declaration of source or composition	Consumer deception
Fraud	Intentional manipulation for economic gain	Loss of trust and safety risks

13.10 Regulatory agencies governing natural and nutraceutical food additives

The safety and quality of natural and nutraceutical food additives are governed by national and international regulatory agencies responsible for establishing standards, conducting risk assessments, and ensuring compliance. These agencies play a critical role in protecting public health by evaluating scientific evidence, setting usage conditions, and monitoring food additive safety across the supply chain.

13.10.1 Codex Alimentarius Commission (FAO/WHO)

The Codex Alimentarius Commission, jointly established by the Food and Agriculture Organization (FAO) and the World Health Organization (WHO), develops international food standards, guidelines, and codes of practice to ensure food safety and facilitate fair trade (Food and Agriculture Organization (FAO), 2017). Codex standards provide the global reference for food additive safety, including definitions, permitted uses, and general principles applicable to both synthetic and natural additives. The Codex General Standard for Food Additives (GSFA) serves as the primary international reference for permitted additives and their conditions of use.

Codex recognizes that natural and nutraceutical food additives must be subjected to the same rigorous safety evaluation as synthetic additives, with decisions based on scientific risk assessment rather than origin.

13.10.2 Joint FAO/WHO Expert Committee on Food Additives

The Joint FAO/WHO Expert Committee on Food Additives (JECFA) serves as the primary international scientific advisory body for the evaluation of food additives. JECFA conducts toxicological assessments, establishes ADI values where appropriate, and provides scientific advice to Codex and national authorities.

For natural and nutraceutical food additives, JECFA evaluates available toxicological data, exposure estimates, and specifications of identity and purity. Where data are insufficient, JECFA may recommend additional studies or restrict usage until safety is adequately demonstrated.

13.10.3 Food and Drug Administration (United States)

In the United States, the Food and Drug Administration regulates food additives under the Federal Food, Drug, and Cosmetic Act. Natural and nutraceutical additives may be regulated as food additives requiring premarket approval or as substances generally recognized as safe (GRAS), depending on the nature of available safety evidence.

The FDA evaluates safety based on scientific data, intended use levels, and estimated dietary exposure, irrespective of whether the additive is of natural or synthetic origin.

13.10.4 European Food Safety Authority (European Union)

The European Food Safety Authority is responsible for the scientific evaluation of food additives in the European Union. EFSA conducts comprehensive risk assessments, including toxicological evaluation, exposure assessment, and reevaluation of previously authorized additives.

Natural and nutraceutical food additives are assessed under the same regulatory framework as other additives, with particular attention to compositional variability and data gaps associated with natural sources.

13.10.5 Food Safety and Standards Authority of India (India)

In India, the Food Safety and Standards Authority of India regulates food additives under the Food Safety and Standards Act and associated regulations. FSSAI specifies permitted additives, usage limits, labeling requirements, and compliance obligations for food businesses.

Natural and nutraceutical additives intended for use in foods must meet safety, quality, and labeling requirements as notified by FSSAI, supported by scientific evidence and regulatory approval.

13.10.6 Other national regulatory agencies

In addition to the agencies described above, several countries maintain national food safety authorities that regulate food additives within their jurisdictions. While regulatory approaches may vary, most national frameworks align with Codex principles and JECFA evaluations, reinforcing international harmonization of food additive safety standards.

13.11 Regulatory compliance for natural and nutraceutical food additives

Regulatory compliance is a fundamental requirement for the safe use of natural and nutraceutical food additives. Compliance encompasses submission of safety documentation, adherence to usage limits, appropriate labeling, and substantiation of any health or nutraceutical claims.

13.11.1 Safety documentation and dossiers

Regulatory authorities require comprehensive safety dossiers for food additives, including information on identity, manufacturing process, specifications, toxicological data, and intended use. For natural additives, additional emphasis is placed on source characterization and compositional consistency.

13.11.2 Acceptable daily intake

The concept of ADI, established through scientific risk assessment, represents the amount of a substance that can be consumed daily over a lifetime without appreciable health risk. ADI values, where established, form the basis for setting usage limits in foods and assessing cumulative exposure. Cumulative exposure from multiple dietary sources must be considered when evaluating safety.

13.11.3 Usage limits and conditions of use

Regulatory authorities specify maximum permitted levels and conditions of use for food additives to ensure consumer safety. These limits are determined based on toxicological evaluation, dietary exposure assessment, and technological need.

13.11.4 Labeling and health/nutraceutical claims

Accurate labeling is essential for regulatory compliance and consumer transparency. Labels must correctly declare the presence of food additives and comply with regulations governing health and nutraceutical claims. Claims must be supported by scientific evidence and must not mislead consumers.

13.12 Risk assessment of natural and nutraceutical food additives

Risk assessment provides the scientific basis for regulatory decision-making and safety management of food additives. Risk assessment follows four key steps: hazard identification, hazard characterization (dose–response), exposure assessment, and risk characterization.

For natural and nutraceutical food additives, hazard identification includes evaluation of intrinsic toxicity, contaminants, and degradation products. Exposure assessment considers intended use levels, consumption patterns, and vulnerable population groups. Risk–benefit evaluation may also be considered where additives provide functional or nutritional advantages, provided safety is not compromised.

Preventive measures such as good manufacturing practices, quality control systems, and postmarket monitoring contribute to effective risk management.

For many natural bioactive compounds, limitations in long-term toxicological data necessitate a precautionary and adaptive risk assessment approach, particularly for bioactive entering food systems through nontraditional or concentrated formulations.

The integrated approach to safety management, quality control, regulatory oversight, and risk assessment for natural and nutraceutical food additives is summarized in Fig. 13.2.

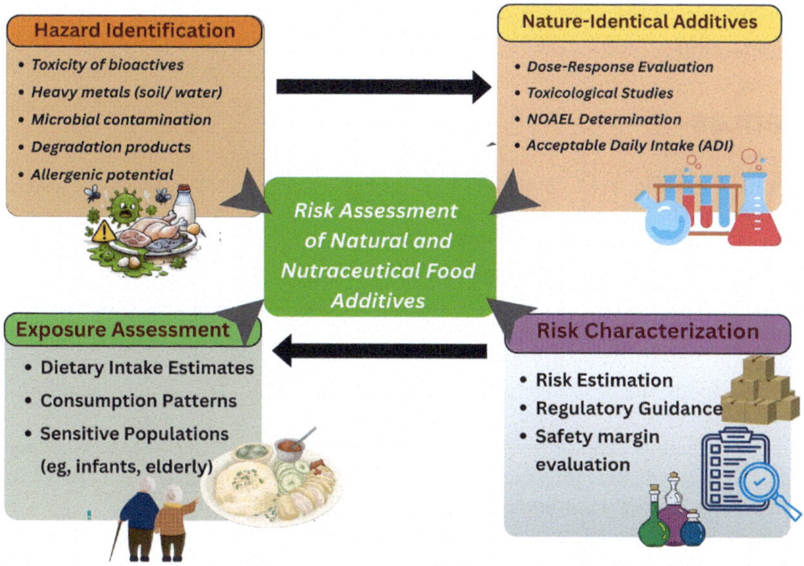

Figure 13.2 Risk assessment framework for natural and nutraceutical food additives. Scientific risk assessment process for natural and nutraceutical food additives involving hazard identification, hazard characterization, exposure assessment, and risk characterization.

13.13 Conclusion and future research perspectives

Natural and nutraceutical food additives offer significant potential for improving food quality and meeting consumer demand for health-oriented products. However, their safe application requires rigorous evaluation, robust quality control, and strict regulatory compliance.

Future research should focus on improving toxicological data for underutilized bioactive, enhancing standardization and stability, and strengthening risk assessment methodologies. Integrating scientific research with regulatory frameworks will be essential to ensure that natural and nutraceutical food additives are both safe and effective in modern food systems. The integration of analytical chemistry, toxicology, and regulatory science is essential for ensuring the safety of natural and nutraceutical food additives.

AI disclosure

During the preparation of this work, the authors used OpenAI and Copilot for the preparation of this manuscript. ChatGPT (OpenAI) was utilized for assistance in refining and organizing the text. Microsoft Copilot was used to generate illustrative images for conceptual visualization. After using this tool/service, the authors reviewed and edited the content as needed and take full responsibility for the content of the publication.

References

Branen, A.L., Davidson, P.M., Salminen, S., Thorngate, J.H., 2002. Food additives, 2nd ed. New York, Marcel Dekker Inc.

Codex Alimentarius Commission, 2023. General Standard for Food Additives (GSFA), CXS 192-1995 (latest revision). Rome, FAO/WHO.

Damodaran, S., Parkin, K.L., Fennema, O.R., 2007. Fennema's food chemistry, 4th ed. Boca Raton, CRC Press.

European Food Safety Authority (EFSA), 2011. Scientific opinion on the evaluation of allergenic foods and food ingredients. EFSA Journal 9 (5), 2187. Available at https://doi.org/10.2903/j.efsa.2011.2187.

European Food Safety Authority (EFSA), 2012. Guidance on the safety assessment of botanicals and botanical preparations intended for use as food supplements. EFSA Journal 10 (5), 2663. Available at https://doi.org/10.2903/j.efsa.2012.2663.

Food and Agriculture Organization (FAO), 2017. The future of food and agriculture—Trends and challenges. FAO, Rome. Available at https://www.fao.org/3/i6583e/i6583e.pdf.

Food and Drug Administration (FDA) (2022). *Food additives status list*. U.S. Food and Drug Administration, Silver Spring, MD. Available at https://www.fda.gov/food/food-additives-petitions/food-additives-status-list.

Joint FAO/WHO Expert Committee on Food Additives (JECFA) (2006). *Combined compendium of food additive specifications*. FAO, Rome.

Joint FAO/WHO Expert Committee on Food Additives (JECFA), 2019. Evaluation of certain food additives. Geneva, WHO Press.

Rahman, M.S., 2007. Handbook of food preservation, 2nd ed. Boca Raton, CRC Press.

Shahidi, F. (2012). Nutraceuticals and functional foods: Whole versus processed foods. Boca Raton: CRC Press.

Shahidi, F., Ambigaipalan, P., 2015. Phenolics and polyphenolics in foods, beverages and spices: Antioxidant activity and health effects—A review. Journal of Functional Foods 18, 820–897. Available at https://doi.org/10.1016/j.jff.2015.06.018.

World Health Organisation (WHO), Food and Agriculture Organisation (FAO), 1987. Principles for the Safety Assessment of Food Additives and Contaminants in Food. WHO Environmental Health Criteria 70, Geneva.

Consumer trends and market insights of nutraceutical food additives

14

Yamuna Mannava[1], *Karthik Maniam*[2], *and Umalatha*[3]
[1]SRK Nutritionals, Ronkonkoma, NY, United States, [2]SRK Nutritionals, LLC, Ronkonkoma, NY, United States, [3]Department of Food Science and Technology, Padmashree Institute of Management and Sciences, Bengaluru, Karnataka, India

14.1 Introduction

Global food consumption patterns have undergone substantial changes due to urbanization, demographic transitions, and lifestyle modifications. These changes have been accompanied by an increased prevalence of diet-related health conditions, prompting consumers to seek foods that support long-term well-being in addition to meeting basic nutritional needs (Puri et al., 2022).

Nutraceutical food additives may be defined as food additives that not only perform technological functions such as preservation, stabilization, or flavor enhancement, but also provide additional physiological or health benefits due to the presence of bioactive compounds. These additives may be broadly classified into functional categories such as natural preservatives, colorants, flavor enhancers, emulsifiers, stabilizers, and bioactive fortifying agents.

Within this evolving context, nutraceutical food additives have emerged as strategically important ingredients that combine technological functionality with perceived health benefits. Unlike conventional additives, nutraceutical-based additives are increasingly positioned as value-enhancing components that align with preventive nutrition approaches and consumer expectations for natural and health-supportive foods (Puri et al., 2022).

From a market perspective, nutraceutical food additives occupy a unique position at the intersection of food technology, nutrition science, and consumer behavior. Understanding consumer trends and market drivers is therefore essential for guiding product development, regulatory strategies, and innovation pathways.

14.2 Evolution of consumer preferences

Consumer attitudes toward food additives have shifted markedly over the past two decades. Earlier acceptance of synthetic additives for convenience and shelf-life extension has gradually given way to increased scrutiny of ingredient composition and processing methods. Consumers now exhibit a strong preference for ingredients perceived as natural, minimally processed, and familiar (Asioli et al., 2017).

Nutraceutical Applications in Food Processing and Preservation. DOI: https://doi.org/10.1016/B978-0-443-45784-5.00001-0

This shift is closely linked to the growing adoption of a preventive health mindset, where food is viewed as a tool for maintaining health rather than merely satisfying hunger or sensory pleasure. Functional foods incorporating nutraceutical additives are therefore increasingly favored, as they allow consumers to integrate health-promoting components into everyday diets without major behavioral change (Bigliardi and Galati, 2013).

Research on consumer trust further indicates that familiarity and perceived naturalness of ingredients significantly influence acceptance, often more strongly than technical efficacy alone (Verbeke, 2015).

14.3 Drivers of market growth

The growth of the nutraceutical food additive market is driven by multiple interrelated factors. Lifestyle-related health concerns such as obesity, metabolic disorders, cardiovascular disease, and digestive health issues have heightened consumer demand for foods with functional benefits (Puri et al., 2022).

In parallel, increased access to nutrition-related information through digital media and scientific communication has improved public awareness of bioactive food components. Peer-reviewed research and regulatory endorsements have strengthened the scientific credibility of nutraceutical ingredients, encouraging their incorporation into mainstream food products (Bigliardi and Galati, 2013).

Additionally, the rapid expansion of the functional food sector has created favorable market conditions for nutraceutical additives by enabling product differentiation and value-based marketing.

Global market assessments indicate steady growth in the nutraceutical and functional food sector, driven by consumer demand for health-oriented products. Market analyses indicate that functional and nutraceutical food categories account for a significant share of new product launches in developed markets, particularly in beverages, dairy, and bakery segments. While regional growth rates vary, Asia-Pacific and North America are frequently identified as key growth regions due to population size, consumer awareness, and innovation intensity (Bigliardi and Galati, 2013)(Puri et al., 2022).

As market growth accelerates in response to these drivers, consumer attention has increasingly shifted toward ingredient transparency and natural formulation practices, giving rise to the clean-label movement (Table 14.1).

14.4 Clean-label and natural trends

The clean-label movement represents one of the most influential consumer-driven trends affecting food additive selection. Clean-label foods are typically characterized by short ingredient lists, recognizable terminology, and avoidance of synthetic additives perceived as artificial or unnecessary (Asioli et al., 2017).

Nutraceutical food additives align strongly with clean-label expectations because many are derived from plant, microbial, or traditional sources and can be communicated as functional and health-supportive ingredients. Transparency,

Table 14.1 Key drivers influencing consumer demand for nutraceutical food additives.

Driver	Description	Implication for nutraceutical food additives
Lifestyle-related health concerns	Rising prevalence of obesity, metabolic disorders, and digestive health issues	Increased demand for additives associated with health support beyond basic nutrition
Preventive health mindset	Shift from reactive to preventive approaches to health management	Preference for foods containing nutraceutical ingredients integrated into daily diets
Nutrition and wellness awareness	Greater access to nutrition information through scientific communication and digital media	Improved consumer understanding and acceptance of nutraceutical-based ingredients
Clean-label expectations	Desire for natural, minimally processed, and recognizable ingredients	Replacement of synthetic additives with nutraceutical alternatives
Expansion of functional food sector	Growth of functional and value-added food categories	Wider application of nutraceutical additives across food matrices

Source: Adapted from Asioli et al. (2017); Bigliardi & Galati (2013); and Puri et al. (2022).

traceability, and clear labeling have become critical determinants of consumer trust, particularly in health-oriented food categories.

Studies on consumer behavior indicate that transparency in ingredient sourcing and labeling significantly enhances acceptance of functional ingredients, even when scientific understanding among consumers is limited (Siegrist and Hartmann, 2020) (Conceptual framework based on consumer behavior and innovation models (Fig. 14.1) (Asioli et al., 2017; Bigliardi and Galati, 2013)).

14.5 Consumer perception of nutraceutical benefits

Consumers increasingly associate nutraceutical food additives with health benefits that extend beyond basic nutrition. Perceived benefits commonly include immune support, digestive health improvement, and metabolic regulation, particularly when nutraceuticals are incorporated into commonly consumed foods such as beverages, dairy products, and bakery items (Puri et al., 2022).

Trust in traditional and natural ingredients also plays a significant role in shaping perception. Ingredients with historical or cultural relevance, such as plant extracts and fermentation-derived compounds, are often regarded as safer and more beneficial, especially when supported by scientific validation and regulatory oversight (Asioli et al., 2017; Verbeke, 2015).

Figure 14.1 Transition from conventional synthetic additives to nutraceutical-based food additives. Illustration of the clean-label transition from conventional synthetic food additives toward nutraceutical-based additives driven by consumer demand for natural, transparent, and health-oriented ingredients.

14.6 Popular categories of nutraceutical food additives

Several categories of nutraceutical food additives have gained prominence due to both consumer acceptance and technological applicability. Natural preservatives are widely used for oxidative and microbial control, while natural colorants are valued for visual appeal and perceived safety.

Natural flavors and flavor enhancers support sensory acceptance, whereas natural sweeteners enable sugar reduction strategies without compromising taste. Natural gums and stabilizers derived from plant and microbial sources are increasingly favored for improving texture and stability while meeting clean-label expectations (Bigliardi and Galati, 2013).

Collectively, these categories reflect a shift toward multifunctional additives that satisfy both technological and consumer-driven requirements.

14.7 Application-wise market trends

Market adoption of nutraceutical food additives varies across food categories. Beverages and functional drinks represent a major application area due to consumer demand for convenience combined with health benefits. Bakery and confectionery products increasingly incorporate nutraceutical additives to enhance shelf life and nutritional positioning.

Dairy and plant-based products serve as key platforms for nutraceutical incorporation, particularly for fibers, probiotics, and bioactive compounds. Nutritional foods and supplements further reinforce consumer familiarity with nutraceutical ingredients, facilitating acceptance across broader food categories (Puri et al., 2022) (Fig. 14.2).

14.8 Regional market trends

Regional differences strongly influence consumer trends and market development. In the Asia-Pacific region, traditional knowledge and herbal ingredients play an important role in shaping consumer trust. In Europe, clean-label demand and regulatory frameworks significantly influence product formulation and marketing strategies (Asioli et al., 2017).

North American markets emphasize functional and fortified foods, supported by high consumer awareness and innovation-driven industry practices. Emerging markets face challenges related to affordability and awareness, but demonstrate

Figure 14.2 Applications of nutraceutical food additives in the food industry. Major food categories incorporating nutraceutical food additives, including functional beverages, dairy products, bakery items, plant-based foods, cereals, and convenience foods.

Table 14.2 Regional consumer trends and challenges in the nutraceutical food additive market.

Region	Dominant consumer trends	Key challenges
Asia-Pacific	Strong trust in traditional, herbal, and plant-based ingredients	Variability in regulatory harmonization and affordability
Europe	High demand for clean-label products and strict regulatory compliance	Stringent health claim regulations and formulation constraints
North America	Preference for functional and fortified foods with scientific backing	Consumer skepticism toward exaggerated health claims
Emerging markets	Growing awareness of nutrition and wellness	Price sensitivity and limited consumer education

Source: Adapted from Asioli et al. (2017); Bigliardi & Galati (2013); and Puri et al. (2022).

growing interest as nutrition education and accessibility improve (Puri et al., 2022) (Table 14.2).

14.9 Consumer challenges and concerns

Despite positive perceptions, several challenges affect consumer acceptance of nutraceutical food additives. Price sensitivity remains a key concern, particularly in cost-conscious markets. Confusion over health claims and limited understanding of scientific validation can also undermine trust.

Sensory acceptance and shelf-life limitations represent additional barriers, as nutraceutical additives may influence taste, texture, or stability if not properly formulated (Puri et al., 2022).

14.10 Role of regulation and consumer trust

Regulatory oversight plays a critical role in shaping consumer confidence in nutraceutical food additives. Clear labeling, validated health claims, and compliance with food safety authorities enhance transparency and trust.

Consumers are more likely to accept nutraceutical additives when products are regulated and approved by recognized authorities such as EFSA, FDA, and national food safety agencies, reinforcing the importance of governance in market development (Puri et al., 2022) (Table 14.3).

14.11 Industry response and innovation

In response to evolving consumer expectations, the food industry has increasingly invested in reformulation strategies that incorporate nutraceutical additives while

Table 14.3 Consumer challenges and corresponding industry responses in the nutraceutical food additive market.

Consumer challenge	Underlying concern	Industry response and strategy
Price sensitivity	Higher cost of nutraceutical ingredients compared to conventional additives	Reformulation optimization and scalable sourcing strategies to balance cost and functionality
Confusion over health claims	Difficulty in understanding scientific validity of nutraceutical benefits	Clearer labeling, evidence-based claims, and regulatory compliance
Sensory acceptance	Changes in taste, texture, or appearance due to natural additives	Improved formulation techniques and sensory optimization
Shelf-life limitations	Stability issues associated with natural bioactive compounds	Advanced stabilization, encapsulation, and processing technologies
Trust and credibility	Skepticism toward unverified or exaggerated claims	Third-party certification, transparent communication, and regulatory oversight

maintaining sensory quality and safety. Advances in extraction, stabilization, and processing technologies have improved the feasibility and consistency of natural bioactive ingredients.

Sustainable sourcing and environmentally responsible processing have also gained importance, aligning nutraceutical innovation with broader sustainability and corporate responsibility goals (Bigliardi and Galati, 2013).

These industry responses reflect a direct alignment between evolving consumer expectations and product development strategies, demonstrating how market demand influences innovation pathways for nutraceutical food additives (Fig. 14.3).

14.12 Future market outlook

Future growth of nutraceutical food additives is expected to be driven by personalized nutrition approaches, integration of traditional knowledge with scientific validation, and increased emphasis on research-backed health claims. As consumers demand greater transparency and evidence-based functionality, nutraceutical additives are likely to play an increasingly central role in modern food systems (Puri et al., 2022) (Puri et al., 2022).

Future market success of nutraceutical food additives will increasingly depend on the ability to balance scientific substantiation, regulatory compliance, sensory quality, and affordability within diverse consumer segments.

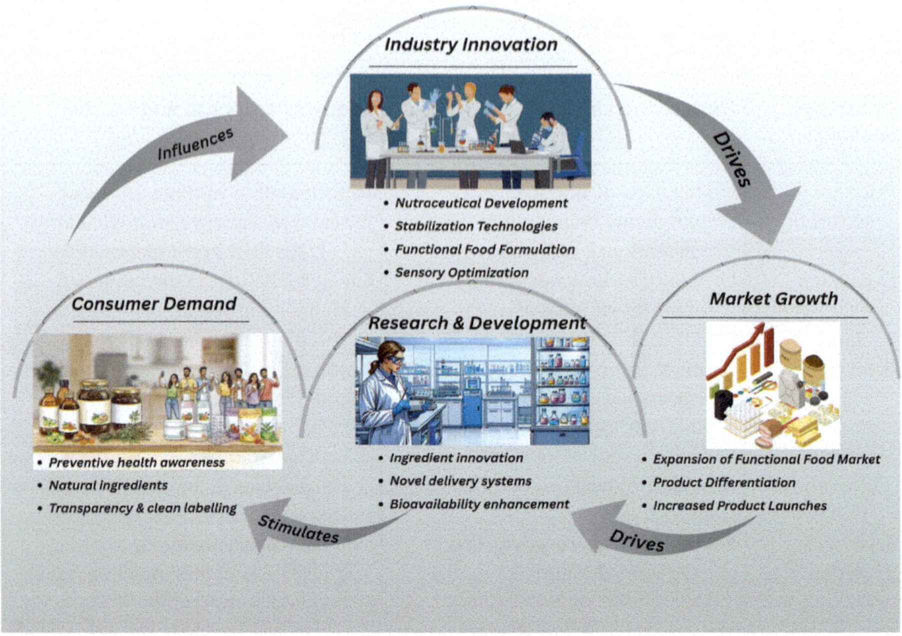

Figure 14.3 Consumer demand–industry innovation–market growth feedback loop for nutraceutical food additives. The figure illustrates the dynamic interaction between evolving consumer preferences, industry-led innovation and reformulation strategies, and market growth of nutraceutical food additives, highlighting a feedback loop that continuously shapes product development and market direction.

AI disclosure

During the preparation of this work, the authors used OpenAI and Copilot for the preparation of this manuscript. ChatGPT (OpenAI) was utilized for assistance in refining and organizing the text. Microsoft Copilot was used to generate illustrative images for conceptual visualization. After using this tool/service, the authors reviewed and edited the content as needed and take full responsibility for the content of the publication.

References

Asioli, D., Aschemann-Witzel, J., Caputo, V., Vecchio, R., Annunziata, A., Næs, T., et al., 2017. Making sense of the "clean label" trends: A review of consumer food choice behavior and industry implications. Food Research International 99, 58–71. Available at https://doi.org/10.1016/j.foodres.2017.07.022.

Bigliardi, B., Galati, F., 2013. Innovation trends in the food industry: The case of functional foods. Trends in Food Science & Technology 31 (2), 118–129. Available at https://doi.org/10.1016/j.tifs.2013.03.006.

Puri, V., Nagpal, M., Singh, I., Singh, M., Dhingra, G.A., Huanbutta, K., et al., 2022. A Comprehensive review on nutraceuticals: Therapy support and formulation challenges. Nutrients 14 (21), 4637. Available at https://doi.org/10.3390/nu14214637.

Siegrist, M., Hartmann, C., 2020. Consumer acceptance of novel food technologies. Nature Food 1 (6), 343–350. Available at https://doi.org/10.1038/s43016-020-0094-x, https://www.nature.com/natfood/.

Verbeke, W., 2015. Consumer acceptance of functional foods: Socio-demographic, cognitive and attitudinal determinants., Newly added (supporting consumer & trust dimensions). Food Quality and Preference 41, 1–9. Available at https://doi.org/10.1016/j.foodqual.2014.11.003.

The future of nutraceuticals in food processing

15

M. Anuradha[1] *and S. Balasubramanya*[2]
[1]Padmashree Institute of Management and Sciences, Bengaluru, Karnataka, India,
[2]Association for Biotechnology Led Enterprises (ABLE), Bengaluru, Karnataka, India

15.1 Introduction

Food sector across the globe is undergoing a dramatic transformation. Earlier days people were primarily focused with producing sufficient food; they are now increasingly concerned about securing nutrition, wellness, and long-term health benefits. This paradigm shift reflects that the foods consumed every day are not only sources of energy, but powerful determinants of physical resilience, metabolic health, and disease prevention. This has multidimensionally influenced every allied segment of food industry; especially nutraceuticals have emerged as a vital bridge between food and preventive medicine, redefining its role in food processing and food sector. This transition reflects the growing recognition that diet quality plays a decisive role in long-term health outcomes and disease prevention and that food systems must evolve to deliver not only calories but also functional nutrition (Granato et al., 2020; Martirosyan and Singh, 2015).

Traditionally, nutraceuticals and herbals were perceived mainly as supplements in preventive healthcare segment. They were marketed as OTC products and consumed as capsules, powders, or tonics. Post-COVID era witnessed a giant leap in usage of nutraceuticals. Recent global assessments also indicate that consumer demand for nutraceutical-enriched foods has increased substantially as individuals seek dietary approaches that support immune resilience, metabolic health, and long-term wellness . However, the seamless integration into food processing plays an important role and well-deserved status to nutraceutical sector. There is global concern over processed foods. However, for ensuring food security and farmer livelihoods, it is imperative to process for extended preservation and usage. Therefore, the emphasis must be focused toward healthy and responsible food processing practices. This clearly demonstrated the future of nutraceuticals and transition from episodic supplementation to continuous nourishment. Hence in this modern era, food processing has become strategic platform. It is not just a tool for preservation and convenience but a weapon for preventive nutrition.

Rising prevalence of lifestyle-related disorders, ageing populations, and the growing burden of chronic diseases increased the pressure on global health systems. These challenges cannot be tackled only by modern healthcare interventions. They need wide array of sustainable dietary solutions that are accessible, affordable, and culturally acceptable. Modern food processing technologies depend on various processing methods and usage of a wide spectrum of food additives both synthetic

Nutraceutical Applications in Food Processing and Preservation. DOI: https://doi.org/10.1016/B978-0-443-45784-5.00016-2

and natural. To have cost benefit, efficiency and easy sourcing synthetic additives prevail universally in the processing industry. Increasing awareness, evolving technologies in nutraceutical industry offer sustainable solutions. Thus, nutraceutical food additives, once embedded, provide dual benefits of providing well-being and processed qualities.

The future of nutraceuticals in food processing is thus significant with respect to food and nutrition security. Hence, food processing is no longer a postharvest inevitability, but a health-enabling technology. This concept opens a new gate through which biodiversity, traditional knowledge, nutraceutical facts, and modern science converge to create foods that are safe, stable, functional, and nutritionally meaningful.

This book presents a visionary perspective on how advances in nutraceutical science can contribute to safer, healthier food processing systems. Though there are challenges in optimization, ingredient sourcing, extraction, purification technologies, delivery systems, still the future of convergence of nutraceuticals and food processing segments is inevitable and invaluable, which can modulate the processing innovations and reshape the nutraceutical landscape. It envisions a future in which nutraceutical additives address many existing challenges of food processing, thus providing holistic food serving as preventive nutrition and sustainable well-being.

15.2 Food processing in the nutraceutical era

The integration of nutraceuticals by replacing the existing harmful synthetic additives redefines the objectives and methodologies of food processing. Conventional food processing is basically designed to address shelf life, texture, safety, stability, and sensory quality. These are essential goals; however, preventive healthcare era demands a more embracing vision, where processing is also expected to enhance the safety and deliver health promoting products to the consumers.

In this new consumer-centric era, food processing can evolve from a preservation-centered technology to a function-driven sector. Ingredients or food additives are no longer selected merely for their functionality, but also for their health benefits. Processing methods are being designed not only to achieve marketable food matrices but also to preserve the nutritional integrity. This results in designer functional foods, which actively contribute to metabolic balance, immune resilience, and wellness. Functional foods formulated with nutraceutical ingredients have demonstrated the ability to contribute to improved physiological functions, immune modulation, and metabolic balance when incorporated into daily diets (Shahidi, Ambigaipalan, 2015; Granato et al., 2020).

The concept of designing foods for function and nutraceutical benefits is an age-old phenomenon. However, industrial research and cost reduction factors contributed to introduction of additives that are harmful. The accumulated data about harmful ingredients used as food additives is mind boggling and many additives that are used are off the shelves as soon as there are reports and scientific validation. For example, the pink color *Rhodamine B* used to color cotton candy is banned from use in food products because it is unsafe for consumption and has been linked to serious health risks. When talking about food colors, almost all colors are used only for appeal and presentation. These can easily be replaced with an array of natural colors extracted from algae and edible plant parts. These plant- and algae-based colors are derived from flavonoids, carotenoids, betalains, chlorophylls, and phycobiliproteins. They not only function as natural coloring agents but also exhibit antioxidant and immune-supporting properties. Likewise, food processing technologies can be engineered with specific health outcomes in mind, whether related to immunity, gut health, cardiovascular function, cognitive performance, or metabolic regulation. Natural pigments derived from plants, algae, and microbial sources are increasingly recognized as multifunctional ingredients that provide both coloring and antioxidant benefits in food systems (Rodriguez-Amaya, 2019).

Another big domain silently dominating the psyche of obese and diabetic population is sugar. Market is flooded with products with artificial sweeteners. Common artificial sweeteners approved include aspartame, sucralose, and saccharin. However, the health researchers advise caution in their long-term use due to possible associations with metabolic disorders and cardiovascular risk, and IARC (2023)has classified aspartame as possibly carcinogenic to humans. In this context, there are many natural sweeteners like steviol glycosides extracted from *Stevia rebaudiana* and mogrosides from monk fruit that are popular and used in many products. There are challenges in adopting nutraceuticals for various reasons. Steviol glycosides though very stable at low and higher temperatures are associated with after taste. Apart from this, processing parameters such as temperature, pressure, moisture, and oxygen exposure must be carefully controlled to minimize degradation and maximize functional retention. Natural high-intensity sweeteners such as steviol glycosides and mogrosides are increasingly incorporated into food products due to their high sweetness potency and favorable safety profiles when used within recommended regulatory limits (Grembecka, 2015).

Most important aspect to note is that the future of nutraceutical usage depends on various factors such as toxicity studies, extraction methods, solvent residues, safety, stability, and sensory appeal. Nutraceutical integrated foods must meet similar standards of quality and consumer acceptance as traditional products. The success of this approach depends on harmonizing the nutraceutical ingredient functionality with taste, texture, convenience, and feasibility.

15.3 Emerging trends of nutraceuticals for food processing applications

The future progression of nutraceutical empowered food systems depends on the continuous research and sustainable utilization of diverse biosources. Historically conventional nutraceutical ingredients have been derived from a limited range of plants, microbes, and marine organisms; the next generation of nutraceutical ingredients can be drawn from a broader biological landscape.

Underutilized and unexplored plants, marine, and microbial sources represent a promising reservoir of nutraceutical with food applications. Many cereals, millets, legumes, tubers, fruits, and leafy vegetables possess exceptional nutrient density and rich profiles of bioactive compounds. Underutilized and traditional crops are gaining renewed scientific attention as sustainable sources of nutraceutical compounds capable of diversifying food systems and strengthening nutritional security (Liang et al., 2022).

These crops, adapted to local agro-climatic conditions, are often cultivated and require fewer inputs, making them ideal candidates for sustainable nutraceutical additive sourcing. The waste resulted from various industrial segments, like vegetables, fruit processing units, wine, and other allied food sectors, are rich sources of various bioactive compounds. These can be researched for sustainable utilization as food additives. Fruit peels, seed coats, seeds, vegetable waste, pomace, bran, husks, and oilseed cakes contain high concentrations of antioxidants, dietary fiber, polyphenols, and functional lipids. Instead of being discarded or underutilized, these materials are increasingly being valorized through advanced extraction technologies, converting waste streams into high-value functional food ingredients. Agroindustrial by-products are increasingly recognized as valuable sources of polyphenols, dietary fiber, and functional lipids, and their valorization represents an important strategy within circular bioeconomy models (Shahidi, Ambigaipalan, 2015; Mirabella et al., 2014)

Microbial, algal, and marine sources are hidden wealth of nutraceuticals representing another frontier of innovation. These are being harnessed to produce colors, flavors, gelling agents, sweeteners, emulsifiers, bioactive peptides, vitamins, organic acids, and many other bioactive compounds.

These are the emerging trends to use underutilized wide spectrum of biological sources. However, the the future of nutraceutical integration into food systems is inseparably linked to the optimization of extraction, purification, and delivery methodologies. Solubility, stability, pH dependency, and temperature resistance are major challenges where future researchers need to focus. As demand grows for functional ingredients, there is an increasing need for technologies that can efficiently isolate bioactive compounds from natural sources while preserving their structural integrity, biological activity, and safety for food applications. Green extraction technologies have emerged as efficient approaches for recovering bioactive compounds while minimizing solvent use, environmental impact, and degradation of sensitive nutraceutical compounds (Chemat et al., 2017).

Nonsolvent and green extraction technologies such as supercritical fluid extraction, pressurized liquid extraction, ultrasound-assisted extraction, enzyme-assisted extraction, and microwave-assisted extraction are promising and progressively gaining attention. These approaches reduce or eliminate the use of solvents, shorten extraction periods, improve the yield, and retain the quality of the bioactives.

Parallelly, advanced purification systems such as membrane filtration, chromatographic fractionation, and sophisticated distillation technologies enable targeted isolation of nutraceutical fractions from complex biological mixtures, allowing the production of standardized ingredients with defined bioactive profiles.

The most transformative developments are occurring in microencapsulation-, nanoencapsulation-, and nanotechnology-based delivery systems. Many nutraceutical compounds are inherently unstable, poorly soluble, or sensitive to heat, oxygen, and light. Encapsulation technologies protect these compounds from degradation during processing and storage, enable controlled release, and improve bioavailability. Encapsulation technologies are widely explored to improve the stability, solubility, and bioavailability of nutraceutical compounds during food processing and storage (McClements, 2020).

From a processing and preservation perspective, these systems offer multiple advantages like protection from thermal degradation, reduced oxidative deterioration, improved shelf-life stability, and enhanced functional performance within food matrices.

15.4 Advanced processing technologies

While integrating nutraceuticals as a functional ingredient or as an additive, it is important to consider the type of processing technology used. The scenario of food processing technologies is also drastically evolving and undergoing technological renaissance. Minimal and nonthermal processing technologies such as high-pressure processing, pulsed electric fields, cold plasma, and ultrasound preserve nutritional integrity while ensuring safety.

Precision fermentation and bioprocessing strategies enable controlled biosynthesis of vitamins, peptides, antioxidants, and probiotic components. Smart processing systems integrating sensors, automation, and digital control platforms ensure consistency, traceability, and optimization. These technologies enable the creation of foods that are not only safe, but also nutritionally robust and functionally optimized. These advanced processing technologies, novel delivery systems, and nanomethods further enhance the possibility of introducing novel nutraceutical additives into the food systems. Emerging nonthermal technologies such as high-pressure processing, pulsed electric fields, ultrasound, and cold plasma have demonstrated strong potential to preserve bioactive compounds while ensuring microbial safety and product stability (Barba et al., 2017).

15.5 Nutraceuticals and personalized nutrition

The future of food and nutrition is progressively moving toward personalization, recognizing that dietary needs vary according to age, genetics, lifestyle, and metabolic profile. Every product has range of variants targeted toward life cycle and lifestyle nutrition and dietary requirements.

Food processing enables modular formulation systems and adaptive processing technologies that allow the development of convenient foods tailored to specific population groups such as children, elderly individuals, sports persons, pregnant women, lactating mothers, and individuals with metabolic disorders.

This evolution represents a shift from population-wide nutritional guidelines to individual-centric nutrition strategies, where food becomes a personalized health tool rather than a generic commodity. Hence while standardizing any novel nutraceutical ingredient for food processing applications, it is imperative to keep in mind about personalized needs and requirements.

The integration of nutraceuticals in future will reshape the food industry at every level. Food product development is increasingly driven by functionality, clean-label positioning, and multitudinous benefits including convenience and ease of use. Food companies are evolving into nutrition-driven and well-being promoting enterprises. Startups and small enterprises are catalyzing rapid innovation, while established manufacturers are investing in nutraceutical research and technology development. The food industry is transitioning from a commodity model to a health-value model. Advances in nutrigenomics and digital health technologies are enabling personalized nutrition strategies that align dietary recommendations with individual genetic, metabolic, and lifestyle characteristics (Ferguson et al., 2016).

15.6 Public health, food security, and sustainability

Nutraceutical integrated food processing holds pragmatic potential for public health, food security, and environmental sustainability. Functional foods and nutraceutical ingredients support the prevention and management of disorders, diseases, age-related health issues, reduce the healthcare burdens, and promote healthy ageing.

Healthy processing technologies strengthen food security by reducing post-harvest losses, stabilizing seasonal surpluses, and ensuring year-round access to nutritious foods. Sustainability is reinforced through circular bioeconomy models, green extraction, waste valorization, and decentralized value chains. Functional foods enriched with nutraceutical ingredients are increasingly recognized as effective tools in addressing micronutrient deficiencies and reducing the risk of chronic noncommunicable diseases (Granato et al., 2020; FAO, 2013).

15.7 Challenges, ethics, and responsible innovation

Despite immense promises, nutraceutical-enabled food systems must be guided by scientific rigor, ethical responsibility, and social equity. Health claims must be evidence-based, and safety, standardization, and quality control must remain priorities. Lack of cultivation practices, solvent residues, and microbial and heavy metal load need to be critically controlled for this novel and potent segment. Adulteration, seasonal variations, and environmental influences create a major problem for scaling up of any novel ingredient optimized. Responsible innovation, safety trials, and toxicity studies are the need of the hour in such approaches. Affordability, accessibility, and feasibility ensure that nutraceutical ingredients in food processing application will have a bigger success. These are vital to serving public health rather than becoming exclusive premium products.

The future of nutraceuticals will be shaped by enabling policies, regulatory clarity, and workforce development. Education systems must integrate nutraceutical science into food technology curricula. Public–private partnerships and innovation ecosystems will accelerate translation from laboratory research to industrial application.

The future of nutraceutical ingredients focuses toward reimagination of food systems. In this vision, food should become an active contributor to human health. Farms should become sources of nourishment and medicine. Processing plants need to evolve into health production platforms. Food systems need to transform into health ecosystems where agriculture, processing, nutrition science, and public health operate as a unified continuum dedicated to sustaining human well-being. Ensuring scientific validation, regulatory oversight, and transparent communication of health benefits is essential to maintain consumer trust in nutraceutical-enabled food systems.

15.8 Conclusion

The integration of nutraceuticals into food processing marks one of the most significant transformations in the history of food systems. What began as an effort to preserve food for survival has evolved into a science of designing foods for health, longevity, and quality of life. Through advances in sourcing, extraction, purification, processing, and delivery technologies, nutraceuticals are becoming integral components of modern diets. Their future lies not in isolated supplements, but in everyday foods that quietly and consistently support human health (Binns, Lee and Lee, 2018; Barba et al., 2015; FAO, 2013; Liang et al., 2022; Mirabella, Castellani and Sala, 2014).

When guided by interdisciplinary collaboration, responsible innovation, and public health priorities, nutraceutical-enabled food processing has the potential to reshape global nutrition, strengthen food security, and build resilient food systems for generations to come. Food is no longer merely sustenance. It is nourishment with purpose. And food processing is the science that makes it possible.

AI disclosure

During the preparation of this work, the authors used OpenAI in order to check spellings and grammar while drafting the chapter. After using this tool/service, the authors reviewed and edited the content as needed and take full responsibility for the content of the publication.

References

Barba, F.J., Parniakov, O., Pereira, S.A., Wiktor, A., Grimi, N., Boussetta, N., et al., 2015. Current applications and new opportunities for the use of pulsed electric fields in food science and industry. Food Research International 77, 773–798. Available from: Available at https://doi.org/10.1016/j.foodres.2015.09.015, www.elsevier.com/inca/publications/store/4/2/2/9/7/0.

Binns, C.W., Lee, M.K., Lee, A.H., 2018. Problems and prospects: Public health regulation of dietary supplements. Annual Review of Public Health 39, 403–420. Available at https://doi.org/10.1146/annurev-publhealth-040617-013638. Available from: http://www.annualreviews.org/journal/publhealth.

Chemat, F., Rombaut, N., Sicaire, A.G., Meullemiestre, A., Fabiano-Tixier, A.S., Abert-Vian, M., 2017. Ultrasound assisted extraction of food and natural products. Mechanisms, techniques, combinations, protocols and applications. A review. Ultrasonics Sonochemistry 34, 540–560. Available from: Available at https://doi.org/10.1016/j.ultsonch.2016.06.035, www.elsevier.com/inca/publications/store/5/2/5/4/5/1.

FAO (2013). *Food systems for better nutrition.* Food and Agriculture Organization of the United Nations, Rome.

Ferguson, L.R., De Caterina, R., Görman, U., Allayee, H., Kohlmeier, M., Prasad, C., et al., 2016. Guide and position of the International Society of Nutrigenetics/Nutrigenomics on personalised nutrition. Journal of Nutrigenetics and Nutrigenomics 9, 12–27.

Granato, D., Barba, F.J., Bursać Kovačević, D., Lorenzo, J.M., Cruz, A.G., Putnik, P., 2020. Functional foods: Product Development, technological trends, efficacy testing, and safety. Annual Review of Food Science and Technology 11, 93–118. Available at https://doi.org/10.1146/annurev-food-032519-051708. Available from: http://www.annualreviews.org/journal/food.

Grembecka, M., 2015. Natural sweeteners in a human diet. Roczniki Państwowego Zakładu Higieny 66 (3), 195–202.

IARC (2023) https://www.iarc.who.int/wp-content/uploads/2023/07/Summary_of_findings_Aspartame.pdf

Liang, L., Liu, X., Wang, Q., Liang, L., Liu, X., Wang, Q., 2022. Nutritional and functional properties of underutilized plant foods. Food chemistry 370 (2022), 131025.

Martirosyan, D.M., Singh, J., 2015. A new definition of functional food by FFC. Functional Foods in Health and Disease 5, 209–223.

McClements, D.J., 2020. Delivery systems for nutraceuticals and functional foods. Annual Review of Food Science and Technology 11, 299–322.

Mirabella, N., Castellani, V., Sala, S., 2014. Current options for the valorization of food manufacturing waste: A review. Journal of Cleaner Production 65, 28–41. Available at https://doi.org/10.1016/j.jclepro.2013.10.051.

Rodriguez-Amaya, D.B., 2019. Update on natural food pigments. Food Research International 124, 200–205.

Shahidi, F., Ambigaipalan, P., 2015. Phenolics and polyphenolics in foods. Journal of Functional Foods 18, 820–897.

Index

Note: Page numbers followed by "*f*" and "*t*" refer to figures and tables, respectively.

I

J

K

L

M

Printed and bound by CPI Group (UK) Ltd, Croydon, CR0 4YY
30/09/2026
02225119-0002